The Astronomers' Magic Envelope

An Introduction to Astrophysics Emphasizing General Principles and Orders of Magnitude

Prasenjit Saha

University of Zurich

Paul A. Taylor

African Institute for Mathematical Sciences

OXFORD
UNIVERSITY PRESS

Great Clarendon Street, Oxford, OX2 6DP,
United Kingdom

Oxford University Press is a department of the University of Oxford.
It furthers the University's objective of excellence in research, scholarship,
and education by publishing worldwide. Oxford is a registered trade mark of
Oxford University Press in the UK and in certain other countries

First Edition published in 2018

Published in the United States of America by Oxford University Press
198 Madison Avenue, New York, NY 10016, United States of America

British Library Cataloguing in Publication Data
Data available

Library of Congress Control Number: 2017958745

ISBN 978–0–19–881646–1 (hbk.)
ISBN 978–0–19–881647–8 (pbk.)

THE ASTRONOMERS' MAGIC ENVELOPE

Preface

Each day since the middle of 1995, NASA's *Astronomy Picture of the Day* has drawn our attention to something other-worldly. Some of the APOD pictures, like the Blue Marble, are unusual views of places we know very well. Some, such as the Andromeda galaxy, are a beautiful detail on objects we can barely see on the sky with our unaided eyes or a small telescope. Some, like the microwave-background sky, are so far from our everyday experience that ordinary language has no words to convey their significance.

Visiting any of these worlds, beyond the very few within light-minutes or -hours, may be the stuff of science fiction. But understanding something of how they work is possible with ordinary science, sometimes quite simple science. That is the subject of this book—astrophysics, which is the part of astronomy dealing with physical explanations, or the branch of physics dealing with other worlds.

This book grew out of a course at the AIMS-SA, the South-African node of the African Institutes for Mathematical Sciences. Most of our students had not seen much astronomy before, and looking at the sky through a small telescope was a first for nearly everyone. But they were familiar with most of the mathematics needed, and enjoyed scientific computing. Above all, they were motivated to educate themselves, never embarrassed about asking for more explanations, and totally unafraid of strange new concepts. A way to build up an introduction to astrophysics then more or less suggested itself.

Working physicists generally, and astrophysicists especially, have great respect for back-of-the-envelope calculations. When, as students, we had first encountered the concept, we thought it meant a crude solution done by someone too lazy to work out something properly. But gradually we came to understand that is not at all what back-of-the-envelope means. It means an abstraction of what is really essential in a problem, to a form where general principles can be applied. The result may be quite close to correct, or it may be ten times too small or large—but the envelope calculation always provides insight, and hints at how a more detailed solution could be found. In the hands of the wisest among our colleagues, a scrap of paper (envelope, napkin) can seem almost magical. So, we thought, why not teach a course explaining how working astrophysicists do this?

Colleagues will recognize the influence of several well-known books. *Astrophysics in a Nutshell* (Maoz, 2016) set the example for us of an introduction to astrophysics today, based on undergraduate physics. The emphasis on general principles and orders of magnitude was inspired foremost by *First Principles of Cosmology* (Linder, 1997). Other texts we consulted for particular topics included *Principles of Stellar Evolution and Nucleosynthesis* (Clayton, 1984), *Black Holes, White Dwarfs and Neutron Stars: The Physics of Compact Objects* (Shapiro and Teukolsky, 1986), the classic *An Introduction to the Study of Stellar Structure* (Chandrasekhar, 1939), and the two books *Cosmology and Astrophysics through Problems* and *An Invitation to Astrophysicss* (Padmanabhan, 1996; Padmanabhan, 2006). *Galactic Dynamics* (Binney and Tremaine, 2008) was an indirect influence; we say little of galaxies in this book, but we have plenty of dynamics. Other enjoyable books are *Astrophysics for Physicists* (Choudhuri, 2010) and *Introduction to Cosmology* (Ryden, 2016).

Naturally, any new book should offer some distinctive features of its own. This one has four, which we hope readers will find interesting and useful.

First, this book is short! It is a tenth as long as *An Introduction to Modern Astrophysics* (Carroll and Ostlie, 2006) and it is minuscule compared to category astrophysics in Wikipedia. To be useful within this length, we have tried to focus on topics that best bring out general principles. As a result many fascinating topics in contemporary astrophysics are regretfully 'beyond the scope of this book'. In particular, we do not cover formation processes, whether of planets, stars, galaxies, or the periodic table, at all. For an initiation into those topics we recommend the only astrophysics book we know of that is even shorter than this one—*Astrophysics: A Very Short Introduction* (Binney, 2016). We assume that readers are familiar with the basic concepts and vocabulary of astronomy—such as for following the explanations in *Astronomy Picture of the Day*—and are comfortable with looking up unfamiliar terms online. For example, the actual definition of a parsec occurs late in the book, but it is assumed that readers basically already know what it means. We do, however, review some essential concepts (Euler angles, Hamiltonians, quantum statistics, and reaction cross-sections), either in the text or as an Appendix.

A second feature of this book is our opportunistic attitude to units. Working astrophysicists often slip in and out of SI units, and sometimes they even define units on the fly. This book will do both, as well. In particular, we introduce Planckian units in Chapter 4 and use them extensively thereafter. In this, we were largely inspired by Brandon Carter's provocative essay *The Significance of Numerical Coincidences in Nature* (Carter, 2007), which showed how Planckian units can bring out the underlying simplicity of many astrophysical processes. Fortunately, our students have also found these units to be useful and informative, overcoming the initial novelty factor with (as in all things) a bit of practice.

Third, it being the 21st century and all, scientific computing (at the level of numerically integrating differential equations and plotting the results) is incorporated alongside other mathematics. Our students worked with the Python ecosystem, but this book does not assume any particular programming language or software. We strongly urge readers to try the computing exercises; some of them will produce pretty pictures, and hopefully all of them will deliver some insight.

The fourth unusual feature of this book is the use of some historical narrative. We will not delve into Newton's love life or Chandrasekhar's religion, fascinating though such topics may be. We will, however, try to get some feeling for the great paradigm shifts that underlie the history of astrophysics, which have their own logic and illuminate where the subject stands today. In many cases these shifts were felt well beyond the scientific community, as well.

Acknowledgements

We gratefully thank the large number of people who have helped and contributed to this project:

- Firstly, AIMS-South Africa for hosting the initial course, from which this text developed, and the staff there who made (and continue making) that a great learning environment;
- Our students, first at AIMS-SA and then at the University of Zurich, for being active and willing participants;
- Donnino Anderhalden, Simone Balmelli, Alice Chau, Jing Chen, Davide Fiacconi, Marina Galvagni, Elena Gavagnin, Rafael Lima, Gaston Mazandu, Irshad Mohammed, Gift Muchatibaya, Andry Rabenantoandro, Dimby Ramarimbahoaka, and Pascal Vecsei, as tutors and assistants who all helped improve the course and its material over the years;
- Raymond Angélil, Yannick Boetzel, Pedro R. Capelo, Philipp Denzel, Stefan Dangel, Daniel D'Orazio, Mladen Ivkovic, Subir Sarkar, Steven Stahler, Liliya Williams and the three anonymous reviewers for OUP, who contributed many ideas and suggestions; and
- The editorial and production team at OUP, especially Sonke Adlung, Harriet Konishi, Charles Lauder, and Lydia Shinoj.

Contents

1 Orbits

As far as we know, every ancient society practiced something recognisable as astronomy. While early astronomers would not have thought in terms of the Earth being in orbit around the Sun and of the Moon being in orbit around the Earth (though there were exceptions, notably Aristarchos of Samos *circa* 250 BCE), we can see from classical calendars that our ancient forbears did closely account for the facts that the Earth's orbital period is not exactly 365 times its spin period, nor is it exactly 12 times the Moon's orbital period (it's closer to $12\frac{7}{19}$, a ratio which was applied in ancient Babylonian and Metonic calendars). Some of the achievements of ancient astronomers are quite startling—for example, the *Antikythera mechanism*, or the observation of *SN 185*. The first was an intricately geared mechanical calculator for predicting the positions of the planets; it was lost in a shipwreck for 2000 years and is still the subject of reconstructive research. The second was the world's first recorded supernova, the impressive explosion of a star at the end of its life. Today, we can see the remnant of that event (for a picture, see[1] APOD `111110`), but modern attention was drawn to it by ancient Chinese astronomers who recorded the explosion as a 'guest star' in 185 CE.

The 17th century, however, brought two completely new developments to astronomy. The first was technological: telescopes were invented, initially for seeing distant things on Earth, but soon turned towards the sky by Galileo and his successors. Even the earliest telescopes had an aperture several times that of the human eye, so that it was as if astronomers' vision had suddenly become ten times sharper and a hundred times more sensitive; further improvements soon followed. The second development was cognitive: no longer satisfied with predicting what would be where on the sky when, astronomers wanted explanations with forces and accelerations. It was the beginning of astro*physics*. To the extent that one can associate the birth of astrophysics with any one individual and event, it would surely be the publication of Newton's *Philosophiæ Naturalis Principia Mathematica* in 1687. Newton's real contribution was not the laws of motion, which were already in use before him. (*Principia* credits the first two laws to Galileo, and the third law to Wren, Wallis, and Huygens.) The idea of a gravitational force varying as inverse distance-squared was also 'in the air' (Newton's frenemy Hooke may have considered it). Working out the consequences, however—using universal gravity to calculate the orbits of planets and moons, and also terrestrial tides—all begins with *Principia*.

[1] This is an abbreviation we will use through this book, referring to NASA's online *Astronomy Picture of the Day* collection. 'APOD 111110' means

`http://apod.nasa.gov/apod/ap111110.html`

which is the particular image for 2011, November 10.

The Astronomer's Magic Envelope. Prasenjit Saha and Paul A. Taylor.

10.1093/oso/9780198816461.001.0001

We therefore begin our introduction to astrophysics by revisiting a topic that *Principia* addresses, and looking at it with our 21st century eyes.[2]

1.1 The Apple and the Earth

In his later years, Newton recounted some of his early reasoning to his first biographer, William Stukeley. According to Stukeley, Newton remarked, 'Why should that apple always descend perpendicularly to the ground... Why should it not go sideways or upwards, but constantly to the earths centre? Assuredly, the reason is, that the earth draws it.' One can speculate about whether the young Newton was really thinking about apples, or whether the apple was just an explanatory device used by the much older Newton. But there is a subtle science question in the quote. According to Newtonian gravity, there is an attractive force pulling between every particle in the apple and every particle in the Earth. How does the apple know the location of the Earth's *centre* so exactly and why does it head *there,* particularly when there is so much other matter around?

It turns out that the integrated gravitational force due to all the particles in a spherical body is equivalent to concentrating the mass at the centre. That is, if you were blindfolded, you wouldn't be able to feel the difference between the pull of a huge rock, a hollow shell, or a tiny pin, as long as they were (1) centred at the same location, (2) of the same mass, and (3) spherically symmetric. When Newton was working, this was a very difficult theorem. But using mathematics developed long after Newton, we can prove it relatively concisely.

First, consider the point-by-point view. In modern notation[3] Newton postulated that, given a particle of mass M at the origin of a coordinate system, another particle at $\mathbf{r}$ and having velocity $\mathbf{v} = \dot{\mathbf{r}}$ will experience an acceleration,

$$\dot{\mathbf{v}} \equiv \ddot{\mathbf{r}} = -\frac{GM}{r^2}\hat{\mathbf{r}}, \tag{1.1}$$

which we may call the gravitational field. Here G is a constant of nature, which reflects how strong the change in motion is. The unit vector $\hat{\mathbf{r}}$ shows that the acceleration is directed along the line connecting M and the particle, with the negative sign meaning that the latter is pulled towards the former. Note that $\mathbf{v}$ does not appear on the right-hand side of the equation, so that the particle's acceleration (that is, its *change* in velocity) is independent of its velocity. In terms of particle properties, the size of the acceleration just depends on its distance r from the mass in an *inverse-square* relationship.

As a more general description, if the mass were not at the system's origin, but at some location $\mathbf{r}_1$, the acceleration would be given by

[2] *Principia,* in original or translation, can be found online, but the 17th-century style of writing equations in words makes the text pretty incomprehensible without expert commentary. *Newton's Principia for the Common Reader* (Chandrasekhar, 1995) explains what Newton wrote, in modern mathematical language.

[3] We will use boldface for vectors, hat symbols for unit vectors, and dots over variables for derivatives with respect to time.

$$\dot{\mathbf{v}} = -GM \frac{\mathbf{r} - \mathbf{r}_1}{|\mathbf{r} - \mathbf{r}_1|^3}, \tag{1.2}$$

instead. Note that there is still an overall inverse-square relationship on the distance $|\mathbf{r} - \mathbf{r}_1|$. We can imagine lines of force emanating spherically outwards from the mass; the denser the lines, the stronger the gravitational field.

Let us now define an integral

$$\oint_S \dot{\mathbf{v}} \cdot d\mathbf{S}, \tag{1.3}$$

which we call the *flux*. This is an integral, over an arbitrary closed surface S, of the field component normal to the surface. We can also think of it as the net number of lines coming out through the surface. If the surface has a wiggly shape, a line of force could come out, go back in, and come out again. But ultimately, every line of force has to come out. If the mass is outside the surface, all lines of force going into the surface have to come out, so there are no net lines of force coming out. This picture suggests that the flux depends only on whether or not the mass is inside, and not on its precise location.

To show this more formally, we invoke Gauss's divergence theorem. The theorem (found in textbooks on calculus or mathematical physics) states that a surface integral of any smooth vector field $\mathbf{F}(\mathbf{r})$ can be replaced by a volume integral, thus

$$\oint_S \mathbf{F} \cdot d\mathbf{S} = \int_V \nabla \cdot \mathbf{F} \, d^3\mathbf{r}, \tag{1.4}$$

where S is a closed surface and V is the volume enclosed by it. The gravitational flux (1.3) can thus be written as

$$\oint_S \dot{\mathbf{v}} \cdot d\mathbf{S} = -GM \int_V \nabla \cdot (\mathbf{r}/r^3) \, d^3\mathbf{r}. \tag{1.5}$$

The origin can be moved to $\mathbf{r}_1$ if desired. Now, by expanding in Cartesian coordinates we can verify that

$$\nabla \cdot (\mathbf{r}/r^3) = 0, \quad \text{except for } r = 0, \tag{1.6}$$

where the divergence becomes singular. Hence, only an integrable singularity at the origin contributes to the volume integral. We conclude that the flux indeed depends only on whether the mass is inside the surface, and not on its precise location.

Now let us consider not one point mass but many masses, or a distribution of mass. Since the gravitational field Eq. (1.2) is linear in the masses, the flux through any closed surface S from a distributed mass will depend only on the mass inside S. Redistributing the enclosed mass will change the field at different points on S, but it will not change the flux. An analogous theorem applies to electric charges, and is known as Gauss's flux law.

With the gravitational version of Gauss's flux law in hand, let us specialize to a spherical mass distribution. The mass need not be homogeneous, just spherically symmetric. Let S be a spherical surface concentric with the mass. Spherical symmetry implies that there is no preferred direction other than radial. So the gravitational field on S must be radial, and moreover it must have the same magnitude everywhere on S. If the field has constant magnitude on S and its integral the flux depends only on the enclosed mass, the field cannot depend on the detailed distribution of the enclosed mass (provided it is spherical). In particular, a point mass at the centre would produce the same gravitational field as any spherical distribution.

Thus, disregarding minor oblateness and unevenness of each body, the Earth indeed draws the apple towards its centre.

Exercise 1.1 If the Earth were hollow, but still spherical, find the gravitational force on an apple inside, in the empty part. Try to use general arguments, avoiding explicit integration.

1.2 Some Simple Orbits

Let us now focus on the motion of a small particle in the vicinity of the mass M. The expression in (1.1) describes how the particle's velocity changes due to the presence of any spherically symmetric object (assuming the particle has a much smaller mass). We refer to (1.1) as an *equation of motion* by which we could track a particle for any given initial position and velocity. We can picture the particle moving around M in a two-dimensional plane, whose two degrees of freedom can be expressed either in terms of Cartesian (x, y) or polar (r, ϕ) coordinate pairs. Before going on to more general solutions later in this chapter, it is interesting to examine some simple cases, namely keeping one variable constant and allowing the other to vary.

Case I. Consider a circular orbit, which has $r = a$ (a constant) and ϕ varying with time. In Cartesian coordinates this is given by

$$(x, y) = a\,(\cos\phi,\ \sin\phi)\,, \tag{1.7a}$$

$$\phi = [GM/a^3]^{1/2}\,t\,, \tag{1.7b}$$

which can be verified by substitution into Eq. (1.1). The grouping of physical constants in (1.7b) essentially describes the angular frequency of the motion, from which the orbital period and speed (= period / circumference) are given by

$$P_{\text{orb}} \equiv 2\pi\sqrt{\frac{a^3}{GM}}\,, \qquad V_{\text{orb}} \equiv \sqrt{\frac{GM}{a}}\,. \tag{1.8}$$

The quantities P_{orb} and V_{orb} will be very useful reference scales for time and velocity as we progress through this chapter and consider more complicated orbits. The meaning of the quantities will be generalized, but the expressions in (1.8) will remain the same.

From the previous section, we know that this orbital solution is valid for *any* mass M having a spherically symmetric distribution that fits within a radius a. Consider any such arrangement of masses within the radius of motion a; these all have the same average density $\bar{\rho} = M/[4\pi a^3/3]$, so that the orbital period can be written as

$$P_{\text{orb}} = \sqrt{\frac{3\pi}{G\bar{\rho}}}\,. \tag{1.9}$$

Similar time scales will be observed later for more general systems.

Case II. Now consider the basic solution of (1.1) with ϕ is constant and r varying with time, which describes motion purely towards and away from the origin along a single line. This is a funny kind of 'radial' orbit, where the particle appears to fall directly to the centre of the mass; the change in r with time does not depend at all on the constant value of ϕ and can be expressed in parametric form:

$$r(\eta) = a\,(1 + \cos\eta)\,, \tag{1.10a}$$

$$t(\eta) = \frac{P_{\text{orb}}}{2\pi}\,(\eta + \sin\eta)\,. \tag{1.10b}$$

While it is not obvious at first (or perhaps even second) sight that this satisfies (1.1), it can be plugged into the equation of motion for verification. Note that there is no explicit expression for $r(t)$.

Say that the orbit starts at $t = \eta = 0$, where $r = 2a$. The particle reaches $r = 0$ when $\eta = \pi$ and $t = P_{\text{orb}}/2$, and then the particle goes back out to $r = 2a$ again at time $t = P_{\text{orb}}$. We can think of it as the limiting case of cometary orbits, which come very close to the Sun from great distances and then leave again over nearly the same return path. The characteristic length a would represent the mean distance from the centre of the sun, around which the comet oscillates. Taking $\bar{\rho}$ as the mean density inside the initial radius of $2a$ gives

$$P_{\text{rad}} = P_{\text{orb}}/2 = \sqrt{\frac{3\pi}{8G\bar{\rho}}}\,. \tag{1.11}$$

The numerical constants in the time scales are a bit different because of the differing radial values, but importantly each is $\propto 1/\sqrt{\bar{\rho}}$.

There is another way to think about the radial solution Eq. (1.10)—one which is not restricted to a single falling body. Imagine a gravitating sphere whose internal support suddenly breaks, causing the sphere to collapse under its own gravity toward its centre. If it was perfectly spherically symmetric to start with, it will remain so as it collapses: every shell will maintain the same enclosed mass even as it falls toward the centre, and that mass's gravitational effect will still be equivalent to that of a central point mass. As a result, the trajectory of each point on the spherical layer would be described by (1.10). As noted earlier, by tracking the outermost layer, the time to fully collapse would just be a function of the average enclosed density. For a uniform initial density ρ, it follows from (1.10) that the time for the whole sphere to collapse to the centre is $P_{\text{orb}}/2$, and this is called the *free-fall time* Remarkably, this time depends on the mass and initial radius only through the average density. The free-fall time scale characterizes both how long it takes for a star to collapse at the end of its life, as well as how long it takes a hydrogen cloud to collapse at the *beginning* of a star's life. In each case we don't consider the oscillations of the orbital solution, because the compression of all the free-falling layers will eventually interrupt the orbital trajectories and place different physics at the forefront.

1.3 The Gravitational Constant

For being such a fundamental feature of the expanses of the Universe, the gravitational constant G is exceedingly difficult to measure in a laboratory. Several different measurement strategies are in modern use, which have determined a value of

$$G = 6.674 \times 10^{-11}\,\mathrm{kg^{-1}\,m^3\,s^{-2}}\,, \tag{1.12}$$

with some uncertainty creeping in at about the fourth digit. This degree of accuracy is much less than most physical constants we will meet in this text.

Interestingly, the combination

$$GM_\odot = 1.3271244 \times 10^{20}\ \mathrm{m^3\ s^{-2}} \tag{1.13}$$

is known much more precisely. As it involves lengths and times but not mass, $GM_\odot$ can be determined entirely from positions, velocities, and accelerations of solar-system orbits, all quantities that can be measured very accurately. In contrast, measuring G in SI units necessitates measuring small forces, which is much more difficult. There is nothing except convention that makes the laboratory gravitational constant in Eq. (1.12) more fundamental than the heliocentric gravitational constant in Eq. (1.13). Indeed, in spacecraft dynamics, it is necessary to change the convention and use $GM_\odot$ and not plain G, because you cannot navigate spacecraft on four significant digits of gravitational constant. In effect, spacecraft dynamics takes the Sun as the standard mass, with the kilogram as a secondary local standard. Other possible forms for the heliocentric gravitational constant are

$$\frac{GM_\odot}{c^2} = 1476.6250\,\mathrm{m} = c \times 4.9254909\,\mu\mathrm{s}\ . \tag{1.14}$$

Yes, the mass of the Sun is known to sub-millimetre and sub-picosecond accuracy, it is just the mass in kilograms that has a large uncertainty.

Since the expressions in Eqs. (1.12–1.14) are very un-memorable, we might look to write G in other more meaningful ways. For example, we can trade seconds and metres in favour of years and astronomical units (au). The definition

$$\mathrm{au} \equiv 1.495978707 \times 10^8\,\mathrm{km} \simeq 499.00\,\text{light-sec}\,, \tag{1.15}$$

or about $8\frac{1}{2}$ light-minutes, corresponds approximately to the average distance between the centres of the Earth and Sun. The value in Eq. (1.13) then becomes

$$GM_\odot \simeq (2\pi)^2\,\mathrm{au^3\,yr^{-2}}\,. \tag{1.16}$$

This equation is simply a statement that the 'strength' of gravity leads to an orbital period of a year at 1 au from the Sun (cf. Eq. (1.8)). It does not correspond exactly to the orbit of the Earth, which is perturbed by the Moon and by other planets, but it is good to five digits.

Still another interpretation of G is as a relation between density and time, as we saw in the previous section. Rewriting (1.12) in density and time units, we have

$$1/\sqrt{G} = 1.075\,\sqrt{\mathrm{g\,/\,cm^3}}\ \text{hours}\,. \tag{1.17}$$

Now the orbital-period relation in Eq. (1.9) tells us immediately that the orbital period of a satellite just above a water/ice planet will be about three hours, independent of the planet's size. To attempt some human-scale intuition, imagine a clock as a gravitating object with its own satellites around it. Since clocks are a few times as dense as water, the orbital period of a low-orbit clock-moon would be roughly one hour. Thus, the point of the minute hand on a clock moves at about its orbital speed. We know from elementary dynamics that one can picture circular orbits arising from centrifugal force cancelling out gravity. So if we can imagine the centrifugal acceleration at the tip of a clock's minute hand, that is how strong the clock's gravity is. More generally, it is a useful principle to remember is that in gravitational phenomena, frequency is proportional to square-root of density.

How did Newton and his contemporaries express the gravitational constant? In the 17th century weights and measures were like currencies today: country-dependent and apt to change. Dimensions related to the Earth, on the other hand, were typically stable and widely understood. Consider the surface gravity on the Earth: $g_\oplus = GM_\oplus/R_\oplus^2$. This can be rewritten in terms of the mean density of the Earth, as

$$G\rho_\oplus = \frac{3}{4\pi}\frac{g_\oplus}{R_\oplus}. \tag{1.18}$$

The right-hand side here was well-measured in the 17th century (it does vary across the Earth, but only by about a percent). In other words, the product $G\rho_\oplus$ was known to good accuracy, and hence measuring $\rho_\oplus$ was equivalent to measuring G. Newton in *Principia* guesstimated the mean density of the Earth as between five and six times water density, and experimenters through the 18th and 19th centuries continued to state their results as measurements of the mean density of the Earth. It is not that they were obsessed with a geophysical parameter and unaware that they were measuring a universal constant; they were simply using the unit conventions of their time.[4]

In summary G is, and always has been, a shape-shifter. On a positive note this malleability means that we should have no qualms about adopting its most convenient expression on a case-by-case basis, even if that means making it disappear (as we will in later applications within this text).

Exercise 1.2 Geostationary satellites are artificial satellites in circular orbit around the Earth, above the equator and with a period of 24 hours, thus remaining stationary on the sky. The orbital radius of a geostationary satellite (measuring from the centre of the Earth) is about 6.6 times the radius of the Earth. Use this fact to estimate the mean density of the Earth.

Exercise 1.3 The length of a year gives another way to estimate the mean density inside the Earth's orbit around the Sun. That formal density, however, is not very interesting. Combine it with the fact, that the Sun has a diameter on the sky of about half a degree, to estimate the actual density of the Sun, in $\mathrm{g\,cm^{-3}}$.

[4]If you are curious about the early history of G measurements, try the 1894 book (which can be found online) entitled *The Mean Density of the Earth* by J. H. Poynting. Poynting—who today is best known for his contributions to electromagnetism—begins unequivocally with, 'The determination of the mean density of the earth may be regarded also as the determination of the complete expression of Newton's law of gravitation'.

Exercise 1.4 Imagine a planet of radius R. On the planet is a pendulum of length l (assumed $l \ll R$). Just above is a low-orbit satellite. Show that the period of the pendulum will be $\sqrt{l/R}$ times the orbital period of the satellite.

1.4 Kepler's Laws

In the early 1600s Kepler showed that observations of planetary paths could be fitted by formulas derived from three postulates now called Kepler's laws:

1. The orbit of a planet traces an ellipse, with the sun located at one focus.
2. A line from the Sun to a planet sweeps out equal areas in equal time intervals.
3. The ratio of orbital time squared, to the orbit's major axis cubed, is the same for all planets.

These laws, which are framed in terms of geometry and average properties, turned out to be consequences of gravitational acceleration due to the Sun. The first major success of the Newtonian description of gravity was explaining planetary orbits and these laws.

The main part of the third law is the shortest to derive, as it is essentially a statement about the scaling properties of the equation of motion in Eq. (1.1). Let's say that we had solved that equation for motion around a particular mass M. If we adjusted the mass, for example by multiplying it by some factor f_M, how would we adjust the other quantities without changing the equation? Scaling the mass and multiplying all lengths by a factor f_r and all times by a factor f_t, yields the acceleration

$$\frac{f_r}{f_t^2}\frac{d^2\mathbf{r}}{dt^2} = -G\frac{f_M M}{f_r^2 r^2}\hat{\mathbf{r}}\,. \tag{1.19}$$

Comparing with the original equation of motion, we see that if

$$f_r^3 = f_M f_t^2\,, \tag{1.20}$$

then the equation remains invariant. This establishes the relation between scale lengths and times in Kepler's third law. Moreover, by including the mass term, it also reiterates the relation between densities and orbital periods (recall: $t \propto 1/\sqrt{\rho}$) discussed in the previous section. It still remains to show that the orbital period depends only on the major axis and not on the eccentricity of the ellipse. That will be done shortly.

Kepler's second law is really just the statement that orbits conserve angular momentum. To see this, let us recall the vector expression for the area of a triangle. If the origin and vectors $\mathbf{c}$ and $\mathbf{d}$ are vertices of a triangle, then $\mathbf{A} = \frac{1}{2}\,\mathbf{c}\times\mathbf{d}$ will be its area, directed normal to the triangle itself. In an infinitesimal time dt, the orbit moves from $\mathbf{r}$ to $\mathbf{r}+\mathbf{v}\,dt$, so that the swept out area $d\mathbf{A}$ would be approximated by the triangle between the origin, $\mathbf{c}=\mathbf{r}$ and $\mathbf{d}=\mathbf{r}+\mathbf{v}\,dt$. The rate of sweeping out an area is then

$$\frac{d\mathbf{A}}{dt} = \frac{\frac{1}{2}(\mathbf{r})\times(\mathbf{r}+\mathbf{v}\,dt)}{dt} = \tfrac{1}{2}\,\mathbf{r}\times\mathbf{v} \equiv \tfrac{1}{2}\,\mathbf{l}\,, \tag{1.21}$$

where we have recognized the specific angular momentum (that is, angular momentum per unit mass) $\mathbf{l}=\mathbf{r}\times\mathbf{v}$. Now consider the rate of change of the sweeping itself, which is expressible in terms of angular momentum and simplifies greatly:

$$d\mathbf{l}/dt = \mathbf{v} \times \mathbf{v} + \mathbf{r} \times \dot{\mathbf{v}} = 0\,. \tag{1.22}$$

The first term is identically zero, and the second term is zero because $\dot{\mathbf{v}}$ is only directed along $\mathbf{r}$ for Newton's expression of gravity (1.1). Hence, $\mathbf{l}$ is conserved for these systems, and the area-sweeping rate is constant, which demonstrates Kepler's second law. In fact the same would be true for *any* system whose acceleration is only directed along $\mathbf{r}$. Moreover, since $\mathbf{l}$ is normal to both $\mathbf{r}$ and $\mathbf{v}$ by construction, conservation of $\mathbf{l}$ implies that the orbit is restricted to a single, unchanging plane.

Next, let us take up Kepler's first law. There are several ways to derive it, but perhaps the most elegant is through a curious-looking quantity this is built from the specific angular momentum and Newton's gravitational acceleration:

$$\mathbf{e} \equiv \frac{\mathbf{v} \times \mathbf{l}}{GM} - \frac{\mathbf{r}}{r}\,. \tag{1.23}$$

This is known as the Laplace–Runge–Lenz vector, but we can just think of it as a vector eccentricity, and it turns out to be conserved ($\dot{\mathbf{e}} = 0$: see Exercise 1.5 below). Looking at the right-hand side, we see that each term is a vector located in the orbital plane, and therefore so must be $\mathbf{e}$ itself. Projecting the eccentricity vector along the orbit's radius $\mathbf{e} \cdot \mathbf{r}$ and rearranging gives

$$r = \frac{l^2/(GM)}{1 + e\cos\phi}\,, \tag{1.24}$$

where ϕ is the angle between $\mathbf{e}$ and $\mathbf{r}$. The family of geometric curves so described was well known to Kepler: they form the edges of slices through a cone known as *conic sections*—see Fig. 1.1. For $e^2 < 1$ the curves are ellipses with the origin being one focus, thus recovering Kepler's first law. The family $e^2 > 1$ are hyperbolas, while a parabola is the transitional case.

To characterize the orbit a little more, let us introduce the scale parameter a, such that $a(1 - e^2) = l^2/(GM)$. In other words, let us rewrite the conic section formula as

$$r = \frac{a(1 - e^2)}{1 + e\cos\phi}\,. \tag{1.25}$$

We note that if $e = 0$, then $r = a$ would be constant (with motions as described in Section 1.2), and we can think of the elliptical cases of $e < 1$ as quantifying perturbations away from a circular orbit. For $e > 1$, the parameter a is negative, but it is still useful for defining the *pericentre* (closest approach) of the orbit; more of this will be discussed below and particularly in Section 1.6.

For an ellipse, a is the semi-major axis, while e is the *eccentricity*, and the maximum and mininum r values are $a(1 \pm e)$. Thus, extending this line of reasoning, $\mathbf{e}$ is a kind of vector eccentricity. Referring back to the definition of V_{orb} in (1.8), for bound orbits we can write

$$l = V_{\mathrm{orb}}\, a\,\sqrt{1 - e^2}\,. \tag{1.26}$$

(But note how the interpretation of a and V_{orb} has changed: for circular orbits these were the radius and speed; now they are constants giving the characteristic scales of

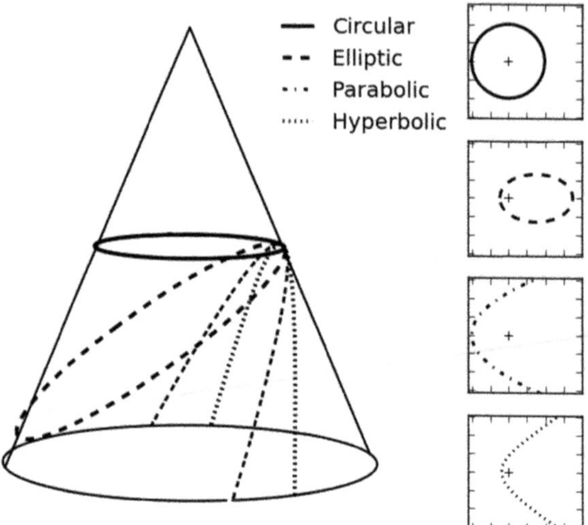

Fig. 1.1 Conic sections, corresponding to bound and unbound orbits. For an upward pointing cone, a horizontal slice intersects as a circle ($e = 0$); a somewhat inclined slice gives an ellipse ($e^2 < 1$); a slice with the same inclination as the slope of the cone gives a parabola ($e^2 = 1$); and a still more inclined slice gives a hyperbola ($e^2 > 1$). Examples of each orbit (looking 'downward' at the orbital plane) are shown at the right.

the varying r and v.) We saw earlier that $l/2$ is the area-sweeping rate. The area of an ellipse is known from geometry to be $\pi a^2\sqrt{1-e^2}$. Dividing area by area-sweeping rate gives $2\pi a/V_{\rm orb}$ as the orbital period. Referring to Eq. (1.8) again, we see that $P_{\rm orb}$ still equals the orbital period, provided a is interpreted as the semi-major axis. This completes the derivation of Kepler's third law.

The eccentricity vector also has an interesting dynamical interpretation. To see it, let us square the definition (1.23). Bearing in mind that $\mathbf{v}$ is perpendicular to $\mathbf{l}$, we obtain

$$e^2 = 1 + \frac{2El^2}{(GM)^2}, \tag{1.27}$$

where

$$E = \frac{v^2}{2} - \frac{GM}{r}, \tag{1.28}$$

which is the energy per unit mass of the orbiting body. Comparing with (1.24), we have

$$E = -\frac{GM}{2a}. \tag{1.29}$$

Since a is constant (recall: it is a scale parameter for any of the orbital cases), E must be constant as well. Hence the total energy is conserved along an orbit. Furthermore, the sign of E gives the character of the orbit. The orbits considered by Kepler had $E < 0$ ($\rightarrow e^2 < 1$) and r remaining finite, which means that they were *bound orbits*: the orbitor could not escape arbitrarily far away. This applies to circular and elliptical orbits. The regime $E > 0$ gives *unbound orbits*, which are hyberbolic, with $e^2 > 1$ and

a being formally negative, and allowing r to become arbitrarily large. In the marginal case of $E = 0$, Eq. (1.25) becomes singular but has a well-defined limit, corresponding to a parabola.

Exercise 1.5 Show that the Laplace–Runge–Lenz vector defined in Eq. (1.23) is conserved, by using standard identities for manipulating vectors. Note that while $\dot{\mathbf{r}} = \mathbf{v}$, $\dot{r}$ is not $|\mathbf{v}|$ but $\mathbf{r} \cdot \mathbf{v}/r$.

Exercise 1.6 Show that the expression (1.25) for a conic section is equivalent to

$$\frac{(x+ea)^2}{a^2} + \frac{y^2}{a^2(1-e^2)} = 1$$

in Cartesian coordinates $(x, y) = (r\cos\phi, r\sin\phi)$.

Exercise 1.7 Plot a few examples of curves given by Eq. (1.25), including both elliptical and hyperbolic cases. One way to do this directly is to plot the Cartesian coordinates $(x, y) = (r\cos\phi, r\sin\phi)$ with r computed as a function of ϕ according to (1.25).

1.5 Kepler's Equation and Time Evolution of Bound Orbits

Kepler's laws do not give the orbital trajectory explicitly as a function of time. Information about the time evolution is nonetheless implicitly present in the equal areas law. Let us now extract that information.

The result of Exercise 1.6 for bound orbits ($e^2 < 1$), expresses the ellipse law in Cartesian form. In words, (i) the centre of the ellipse is displaced from the origin by ae, and (ii) the extent along y is $\sqrt{1-e^2}$ times that along x. These two statements can be turned into a parametric expression for the ellipse:

$$(x, y) = a\left(\cos\eta - e, \sqrt{1-e^2}\sin\eta\right). \tag{1.30}$$

Using the definition of the magnitude radius $r = \sqrt{x^2+y^2}$, the above relation leads to

$$r = a(1 - e\cos\eta). \tag{1.31}$$

The specific angular momentum, being normal to the orbital plane, will have magnitude $l = l_z = x\dot{y} - y\dot{x}$. Comparing this with the expression for l in (1.26), we have

$$x\dot{y} - y\dot{x} = V_{\rm orb}\, a\sqrt{1-e^2}\,. \tag{1.32}$$

Putting in the coordinate values from (1.30) and simplifying yields

$$r\frac{d\eta}{dt} = V_{\rm orb}\,. \tag{1.33}$$

Using (1.31) and integrating (while recalling that $P_{\rm orb} = 2\pi a/V_{\rm orb}$) yields

$$t(\eta) = \frac{P_{\rm orb}}{2\pi}(\eta - e\sin\eta)\,. \tag{1.34}$$

This is known as *Kepler's equation*, and it gives $\eta(t)$ implicitly. Computing $\eta(t)$ requires numerical root-finding.[5] This completes the solution of Keplerian orbits. Looking back at Section 1.2, we see that circles and free falls are limiting cases, $e \to 0$ and $e \to 1$, respectively.

The description of Keplerian orbits fitted the then-known planetery orbits with accuracy close to the observational limits available *circa* 1700, but there were some discrepancies in matching their calculations to observation. Newton and his contemporaries had the notion that the Keplerian orbits around the Sun were not exact descriptions. While Edmond Halley is most famous for identifying mysterious historical apparitions in the sky as a single object (now called Halley's Comet, see APOD 100104, 000805, and 031003) on a highly eccentric orbit that dipped into the inner Solar System every 76 years, he made many contributions to astronomy. One noteworthy one here is that he worked out that the orbit of his own famous comet was not quite Keplerian; he realized that the planets (mainly Jupiter) were perturbing it and then applied the Newtonian theory to calculate those perturbations.

Exercise 1.8 Show that for a Keplerian orbit, the time-averaged value of $1/r$ equals $1/a$.

Exercise 1.9 For unbound orbits, $e^2 > 1$ and $a < 0$. The expression in Exercise 1.6 is still valid, but all the expressions in this section must be modified. Derive the modified versions of the radial solution in Eq. (1.31) and Kepler's equation in Eq. (1.34).

1.6 Unbound Orbits and Gravitational Focusing

Unbound orbits ($E > 0$ and $e > 1$) have an interesting application in the context of planet formation. The Earth and other rocky planets are thought to have coagulated from a large number of small rocky bodies (known as planetesimals). If there are a large number of planetesimals in a certain region, which can collide with each other and sometimes stick together, there is a tendency for a single body (or *protoplanet*) to grow in what is termed an 'oligarchic' fashion—that is, once a particular object grows a bit larger than its neighbours, it races ahead to a much larger size more quickly. There are two reasons for this. First, once a protoplanet starts to grow, it presents a larger area for collisions and therefore has greater likelihood to directly catch more passers-by. Second, the large body has a larger gravitational influence to attract planetesimals to itself, also increasing the probability of collision.

Let us consider a general collision or near-collision between a comparatively large protoplanet, our central object of interest here having mass M, and a much smaller one or planetesimal. Initially, the planetesimal is far away and has some finite velocity which can be written in terms of orbital energy from Eq. (1.28) as $v_\infty = \sqrt{2E}$ (since 'far away' can be translated as $r \approx \infty$). A useful notion here is the so-called impact parameter b, defined as the predicted closest approach if there were no gravity—that is, the distance of the small planetesimal when its straightline trajectory would

[5] As it happens, Kepler's equation also appears in a geocentric system, and for this reason had in fact been studied long before Kepler. In the 9th century Habash al Hasib was already computing numerical solutions.

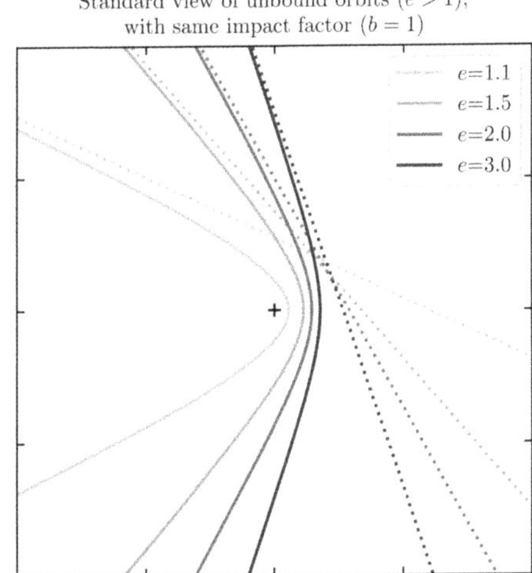

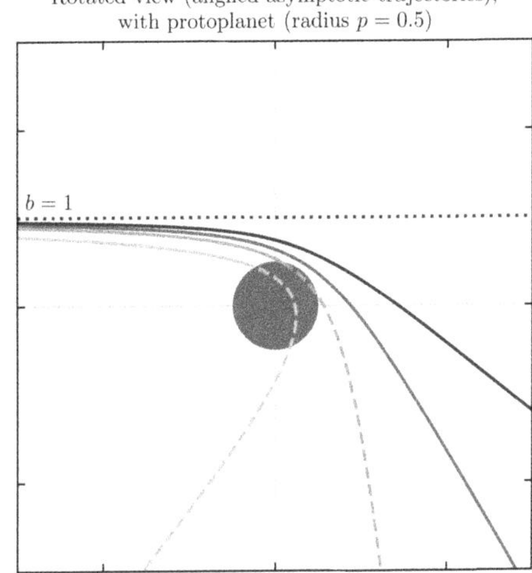

Fig. 1.2 *Left:* Hyperbolic trajectories for unbound orbits (travelling from upper/left) around a central mass at the origin ('+'). Orbits are shown for various eccentricities e (solid lines), along with the hyperbolic asymptote in each case (dotted lines, matched shade). The trajectories have the same impact factor b, i.e, the same minimal distance between the asymptote and central mass. *Right:* the same trajectories are shown rotated anti-clockwise, so that each asymptote lies on the line $y = b = 1$. A protoplanet with radius $p = 0.5$ has been included. Collision (and possible growth of the protoplanet) occurs if the orbit's pericentre is $\leq p$ (dashed lines indicate such cases, showing the theoretical trajectory if no collision had occurred).

be perpendicular to its position vector with the protoplanet. The specific angular momentum is still a constant over the trajectory, so we can calculate its value based on v_∞, which can conveniently be related to the impact parameter: $l = bv_\infty = b\sqrt{2E}$. The expression in Eq. (1.27) for eccentricity then becomes

$$e^2 = 1 + 4\frac{E^2 b^2}{(GM)^2} \cdot \tag{1.35}$$

When $p = a(1 - e)$ is written for the pericentre distance (recalling that $a < 0$ here with $e > 1$), Eq. (1.25) becomes

$$e = 1 + 2\frac{Ep}{GM} \cdot \tag{1.36}$$

Comparing these two equations gives

$$b^2 = p^2\left(1 + \frac{GM}{pE}\right) \equiv p^2(1 + \Theta), \tag{1.37}$$

defining the Safronov parameter $\Theta > 0$.

The interpretation is as follows (see Fig. 1.2). Suppose the protoplanet has radius p. A planetesimal with impact parameter $b \leq p$ will collide directly with it, even without gravity. But Eq. (1.37) quantifies how the gravity of the protoplanet ensures that some planetesimals having even larger impact parameters will still be pulled into collisions. The dimensionless Safronov number is the ratio of the potential energy GM/p at the surface of the protoplanet to the orbital energy E per unit mass of the planetesimal. Alternatively, 'running the video in reverse', we can think of Θ in terms of the local escape velocity $v_{\rm esc}$ of the protoplanet:

$$\Theta = v_{\rm esc}^2 / v_\infty^2 \,. \tag{1.38}$$

As an aside, this discussion also touches on a very modern astronomical topic worth remarking upon. This formation mechanism is the basic reason why Pluto, previously considered a planet (as it is called in an older APOD 990213), was reclassified as a dwarf planet in 2006. One property that a full-fledged planet is defined to have is that it has cleared its neighbourhood of comparable bodies, having either absorbed them, captured them as satellites, or gravitationally influenced their orbits to push them into some other region (i.e., they have somehow 'out-oligarched' everyone else nearby). Dwarf planets like Ceres (APOD 151211, 160204 and the earlier APOD 060821), earlier classified as an asteroid, and Pluto may be larger than their neighbours (Pluto does in fact have moons; see APOD 060624 and APOD 151002), but they have not cleared their neighbourhoods and therefore are not considered to be full planets.

1.7 Two Massive Bodies

So far we have assumed that the orbiting body has negligible mass, effectively producing no appreciable acceleration on the first body and just moving around within the latter's essentially static gravitational field. Let us now relax this assumption and generalize our cases of orbital motion a bit. Consider two spherical bodies with respective mass, position, and velocity: $(m_1, \mathbf{r}_1, \mathbf{v}_1)$ and $(m_2, \mathbf{r}_2, \mathbf{v}_2)$. Let $\mathbf{r}$ and $\mathbf{v}$ be the separation and relative velocity, respectively:

$$\mathbf{r} \equiv \mathbf{r}_2 - \mathbf{r}_1 \,, \qquad \mathbf{v} \equiv \mathbf{v}_2 - \mathbf{v}_1 \,. \tag{1.39}$$

When Newton's laws of motion and gravity are invoked, the forces on the two bodies can now be written as

$$m_2 \dot{\mathbf{v}}_2 = -m_1 \dot{\mathbf{v}}_1 = -\frac{G m_1 m_2}{r^2} \hat{\mathbf{r}} \,. \tag{1.40}$$

It follows that

$$\dot{\mathbf{v}} = -\frac{G M_{\rm tot}}{r^2} \hat{\mathbf{r}} \,, \qquad M_{\rm tot} \equiv m_1 + m_2 \,. \tag{1.41}$$

This equation has the same form as the original equation of motion (1.1). Thus, Kepler's laws and all the other consequences we have derived apply to the new system. In other words, although Kepler's laws were derived for a negligible mass particle in the gravitational field of a central object, the same results apply to any system of two spherical, gravitating bodies with arbitrary masses, *provided* we use the relative coordinates and velocities from Eq. (1.39). Evaluating $\mathbf{r}_1$ and $\mathbf{r}_2$ individually requires one further step, namely consideration of the system centre of mass (or *barycentre,* as commonly termed in gravitational dynamics), as outlined in Exercise 1.10 at the end of this section.

Another useful way of writing the two-body problem is to define the *reduced mass*:

$$M_{\rm red} \equiv \frac{m_1 m_2}{m_1 + m_2} \,. \tag{1.42}$$

If the m_1 and m_2 are very different, the reduced mass is close to the lighter of the two. The equation of motion (1.41) can be written in terms of the reduced mass as

$$M_{\rm red}\dot{\mathbf{v}} = -\frac{G M_{\rm tot} M_{\rm red}}{r^2}\,\hat{\mathbf{r}} \,. \tag{1.43}$$

The formulations in Eq. (1.41) using the total mass, and Eq. (1.43) using the reduced mass, are equivalent. The former is simpler if we are considering just two gravitating bodies, since there is no new quantity to introduce. The formulation with the reduced mass, however, has the advantage that is generalizes directly to other kinds of system. For example, for two electric charges, the right-hand side of Eq. (1.43) changes to an electrostatic interaction, but the left-hand side remains the same.

Exercise 1.10 Show that the centre of mass (or *barycentre*) of two gravitating bodies experiences no acceleration. In a coordinate system with the barycentre as the origin, show further that (i) the bodies are located at

$$\mathbf{r}_1 = -(m_2/M_{\rm tot})\,\mathbf{r}\,, \qquad \mathbf{r}_2 = (m_1/M_{\rm tot})\,\mathbf{r}\,, \tag{1.44}$$

(ii) the characteristic velocities of each individual body are rescaled: $V_{\rm orb\,1} = (m_2/M_{\rm tot})V_{\rm orb}$ and $V_{\rm orb\,2} = (m_1/M_{\rm tot})V_{\rm orb}$, and (iii) the kinetic energies of the two bodies sum to $\frac{1}{2}\,M_{\rm red}\,v^2$.

1.8 Two Bodies from Far Away

Science-fiction authors have long written of planets with two or more suns (Tatooine is one that made it into popular culture). In recent years some circumbinary planets have been discovered, where a planet orbits a close binary-star system. Images of such systems are still science-fictiony (see APOD 160220) but the gravitational field experienced by a circumbinary planet is quite interesting.

Let the masses and locations of the two stars be as in Eq. (1.44) and let $\mathbf{R}$ be the location of the planet, relative to the barycentre of the system. For a basic understanding of the dynamics, let us neglect the mass of the planet. The acceleration of the planet can then be written as

$$\ddot{\mathbf{R}} = -G\,\nabla\, V(\mathbf{R}), \tag{1.45}$$

where

$$V(\mathbf{R}) = -\frac{m_1}{|\mathbf{R}-\mathbf{r}_1|} - \frac{m_2}{|\mathbf{R}-\mathbf{r}_2|} \,. \tag{1.46}$$

Here V is a potential function, conventionally defined such that minus its gradient gives the acceleration of the planet.

Consider the first term in the potential (1.46). Putting aside the minus sign for the moment, we have

$$\frac{m_1}{\left|\mathbf{R} + \frac{m_2}{M}\mathbf{r}\right|} = \frac{m_1}{R}\left(1 + \left(\frac{m_2 r}{MR}\right)^2 + 2\frac{m_2 r}{MR}(\hat{\mathbf{r}}\cdot\hat{\mathbf{R}})\right)^{-1/2}. \tag{1.47}$$

We now assume that the planet is much further away from the stars than they are from each other ($R \gg r$). Accordingly, we expand in powers of r/R and keep only the leading orders.

The bracketed expression in (1.47) has the form

$$(1 + h^2 + 2hx)^{1/2} = 1 - xh + \tfrac{1}{2}(3x^2 - 1)h^2 + O(h^3)\,. \tag{1.48}$$

The little polynomial in x here is well known in mathematical methods and has its own symbol and name[6]

$$P_2(x) \equiv \tfrac{1}{2}(3x^2 - 1)\,. \tag{1.49}$$

Substituting the series (1.48) together with the symbol (1.49) into the potential contribution (1.47) gives

$$\frac{m_1}{|\mathbf{R} - \mathbf{r}_1|} = \frac{m_1}{R}\left(1 - \frac{m_2 r}{MR}(\hat{\mathbf{r}}\cdot\hat{\mathbf{R}}) + \left(\frac{m_2 r}{MR}\right)^2 P_2(\hat{\mathbf{r}}\cdot\hat{\mathbf{R}})\right) + \ldots. \tag{1.50}$$

Putting in the contribution of the second star, we have

$$V(\mathbf{R}) \approx -\frac{M}{R} - M_{\mathrm{red}}\,\frac{r^2}{R^3}\,P_2(\hat{\mathbf{r}}\cdot\hat{\mathbf{R}})\,. \tag{1.51}$$

Thus, the gravitational field seen by a circumbinary planet is as if the total mass were at the barycentre, plus a perturbation smaller by $O(r^2/R^2)$. In the perturbing term, the factor $M_{\mathrm{red}}r^2$ equals the moment of inertia:

$$I = m_1 r_1^2 + m_2 r_2^2\,. \tag{1.52}$$

The P_2 factor depends on the orientations of $\hat{\mathbf{r}}$ and $\hat{\mathbf{R}}$, and is a number between $-\frac{1}{2}$ and 1.

Actually, there is more to this example than circumbinary planets. We have here a simple but incessantly time-varying gravitational field. How does this time-varying information propagate? In classical dynamics, the gravitational field gets updated instantly everywhere, and this is a good-enough approximation for many purposes—but we know from elementary relativity that any information, including about the current gravitational field, can propagate at most at the speed of light. So what really goes on with the gravitational field? We will return to this question towards the end of this book, in Section 8.6.

Exercise 1.11 In the formula (1.51), since $\hat{\mathbf{r}}$ will rotate with the period of the binary, $V(\mathbf{R})$ would appear to be periodic in the same way. But, in fact, $V(\mathbf{R})$ will have half the period of binary. Why is that?

[6]Look up Legendre polynomials, if you are interested further.

1.9 More than Two Bodies

For gravitational systems of $N > 2$ bodies, the orbit equations are a straightforward generalization of the two-body equations:

$$\dot{\mathbf{v}}_i = G \sum_{j \neq i}^{N} m_j \frac{\mathbf{r}_j - \mathbf{r}_i}{|\mathbf{r}_j - \mathbf{r}_i|^3} . \tag{1.53}$$

Unfortunately, the method of *solution* used for two bodies does not generalize at all, even to $N = 3$ bodies. One must then resort to particular situations, approximations, and/or numerical methods.

We will study some the dynamics of three bodies in the next chapter. For larger N-body systems, the use of numerical simulations is standard. Most of these use some form of the leapfrog algorithm, which tracks the particle positions and velocities over snapshots taken at small time intervals Δt, evolving and updating the variables in quick succession, as follows.

$$\begin{aligned} \mathbf{r}_i &\to \mathbf{r}_i + \tfrac{1}{2}\Delta t\, \mathbf{v}_i \\ \mathbf{v}_i &\to \mathbf{v}_i + \Delta t\, G \sum_{j \neq i}^{N} m_j \frac{\mathbf{r}_j - \mathbf{r}_i}{|\mathbf{r}_j - \mathbf{r}_i|^3} \\ \mathbf{r}_i &\to \mathbf{r}_i + \tfrac{1}{2}\Delta t\, \mathbf{v}_i \end{aligned} \tag{1.54}$$

The numerical error goes down as $|\Delta t|^2$ as Δt is reduced.

Exercise 1.12 From Eq. (1.20) it follows that in the two-body problem, rescaling lengths and masses such that densities remain invariant will leave time invariant. Does the same apply to three or more gravitating bodies?

1.10 The Virial Theorem

Notwithstanding the complexities of many-body systems, there is one simple, useful result that applies to any system held together by gravity. For this, we look at a system of many gravitating bodies from a statistical perspective, and get insight from the expectation values of particular quantities. Similarly to the field of statistical mechanics, we trade the ability to look at individual objects for a 'zoomed out' perspective on the system, which still provides very useful physical information.

To derive it, consider the (scalar) quantity obtained by projecting each particle's momentum along its position vector,

$$X \equiv \sum_{i}^{N} m_i\, \dot{\mathbf{r}}_i \cdot \mathbf{r}_i \,, \tag{1.55}$$

where m_i are each of the N masses, which are constants here. Consider the system to be in a steady state, which means that it is isolated and nothing escapes or is added (all trajectories are bounded). Then X itself will be bounded and the long-time average or expectation value $\langle dX/dt \rangle$ will be zero:

$$\begin{aligned} 0 = \left\langle \frac{dX}{dt} \right\rangle &= \left\langle \sum_i m_i \dot{\mathbf{r}}_i^2 + \sum_i m_i \ddot{\mathbf{r}}_i \cdot \mathbf{r}_i \right\rangle \\ &= \left\langle \sum_i m_i \dot{\mathbf{r}}_i^2 \right\rangle + \left\langle \sum_i m_i \ddot{\mathbf{r}}_i \cdot \mathbf{r}_i \right\rangle . \end{aligned} \tag{1.56}$$

In the first term we can recognize the quantities being summed as twice the individual kinetic energy values for each particle; therefore, we identify it as being the expectation value of twice the system's total kinetic energy E_{kin}.

For the second sum, assuming that each particle's acceleration is only due to the gravity of the others, we can substitute in the N-body formula from (1.53), obtaining

$$\left\langle \sum_i m_i \ddot{\mathbf{r}}_i \cdot \mathbf{r}_i \right\rangle = \left\langle \sum_{i,j \neq i} -G m_i m_j \frac{(\mathbf{r}_i - \mathbf{r}_j)}{|\mathbf{r}_i - \mathbf{r}_j|^3} \cdot \mathbf{r}_i \right\rangle . \tag{1.57}$$

This *double* sum encodes the contributions of gravitational attraction between each pair of particles, still projected along a position vector. Each of the N particles is attracted by $N-1$ others, so there are $N^2 - N$ total terms in the summation. In looking to simplify, we can note that the gravitational acceleration of some particle a on another one b, will be equal and opposite to b's acceleration on a; this symmetry is an expression of Newton's third law of motion. Can this help simplify the summation? Well, let's see how the related contributions of two particles might simplify terms in the summation:

$$\begin{aligned} & -G m_a m_b \frac{(\mathbf{r}_a - \mathbf{r}_b) \cdot \mathbf{r}_a}{|\mathbf{r}_a - \mathbf{r}_b|^3} + -G m_b m_a \frac{(\mathbf{r}_b - \mathbf{r}_a) \cdot \mathbf{r}_b}{|\mathbf{r}_b - \mathbf{r}_a|^3} \\ = & -G m_a m_b \frac{(\mathbf{r}_a - \mathbf{r}_b) \cdot (\mathbf{r}_a - \mathbf{r}_b)}{|\mathbf{r}_a - \mathbf{r}_b|^3} \\ = & -\frac{G m_a m_b}{|\mathbf{r}_a - \mathbf{r}_b|} . \end{aligned} \tag{1.58}$$

This has reduced things quite a bit—and as a bonus, this $1/r$ form is also recognizable as that of the gravitional potential energy between two particles. The same simplification happens between any pair of particles, and the summation over all such terms is identifiable as a description of the total gravitational potential energy E_{grav} of the system. Therefore, Eq. (1.56) simplifies to (note that we now have half as many gravitational terms to sum over, since we have made pairs that combined into a single term, as in Eq. (1.58)):

$$\begin{aligned} 0 &= 2 \left\langle \sum_i \tfrac{1}{2} m_i \dot{\mathbf{r}}_i^2 \right\rangle + \left\langle \sum_{i,j>i} -\frac{G m_i m_j}{|\mathbf{r}_i - \mathbf{r}_j|} \right\rangle \\ &= 2 \langle E_{\text{kin}} \rangle + \langle E_{\text{grav}} \rangle . \end{aligned} \tag{1.59}$$

This is known as the *virial theorem*, and states that regardless of the details of a physical system—whatever complicated motions individual constituents make, whether observed or not—we can estimate some important 'bulk' properties of the system as a whole. The virial theorem will appear in another guise in Chapter 5 of this book.

The gravitational N-body equations are a good starting approximation for observing galaxies or clusters of galaxies. In such systems, stars or even entire galaxies, respectively, play the role of the masses m_i. It is usually not possible to measure the individual $\mathbf{r}_i$ and $\dot{\mathbf{r}}_i$ due to their distance, but it is often feasible to measure the projection $|\mathbf{r}_i - \mathbf{r}_j|$ on the sky from imaging as well as the line of sight component of $\dot{\mathbf{r}}$ from spectroscopy. By applying such measured values to a virial expression such as (1.59), along with some model assumptions, one can estimate the total mass of the system. In the 1930s F. Zwicky used this strategy on the Coma cluster of galaxies (see APOD 060321), and the mass came out hundreds of times what would be expected from the starlight. Zwicky then suggested that most of the mass of that cluster is not contained in the bright, shining stars that we can see, but instead it is made up of some kind of 'dark' matter that we don't see directly. Whether dark matter really exists continued to be debated till the 1980s, but the evidence for unseen mass noticeable only from its gravity is now overwhelming. In fact, further studies of both galactic dynamics and cosmology have suggested that dark matter makes up a much larger fraction of the Universe than visible matter. But it remains completely uncertain *what* the dark matter is, or even what all of its exact properties are. This will be discussed further in Chapter 8.

1.11 Connecting to Observables

When studying the gravitational two-body problem thus far, we have taken advantage of the fact that angular momentum conservation confines the orbits to a plane, and written the orbital solution in two dimensions. We should note that, in practice, our viewing angle of systems may not always be conveniently situated to have a 'top-down' view of the orbit, i.e., one that is parallel to the angular momentum vector. Take, for example, the nearest stars after the Sun, known as α Centauri (APOD 110703, 120628, and 160825). This really consists of two stars, α Cen A and B, in Keplerian orbits with $P_{\rm orb} = 80\,{\rm yr}$. (The system also has at least one planet, as well as a third star, but that is much smaller and further away. We are ignoring such minor perturbers.) A coordinate system for observing would have the z axis along the line of sight, and the x, y plane tangent to the sky. The origin can be placed at the centre of mass, but the orbit will be inclined with respect to the x, y plane. Hence, the coordinate system needs to be rotated.

A further complication is that orbits cannot always be observed in three dimensions. The equivalent of stereoscopic vision in astronomy, known as *parallax*, comes from observing the same object from multiple locations on the Earth or different locations along the Earth's orbit. Thus, to measure the unknown distance *to* an object, an observer can use the distance *between* their own locations and the angular shift of the object (via the geometrical rules of triangles). Parallax provides an excellent 3D map of the Solar System. When it comes to α Cen A and B, parallax gives the distance to better than 0.1%, but that error is too large to provide the *difference* in distances to the two stars. In terms of the orbital coordinates with respect to the centre of mass, the depth-like coordinate z is not observable. As for (x, y), what is actually observed is the angular position

$$(\theta_x, \theta_y) \equiv (x/D, y/D)\,, \tag{1.60}$$

where D is the distance from us to the origin of the coordinate system. That distance may or may not be known in advance.

On the good side, the velocity component v_z (the line of sight velocity) is measurable spectroscopically. The spectra of stars have many well-calibrated lines (see APOD 000815 for the Sun's spectrum). If a line known to be intrinsically at wavelength λ is observed shifted to $\lambda_{\rm obs}$,

$$\frac{\lambda_{\rm obs}}{\lambda} = 1 + \frac{v_z}{c} \tag{1.61}$$

gives v_z, where c is the speed of light. The line of sight velocity is commonly called the *radial velocity*—we emphasize that in this case the velocity is 'radial' with respect to the observer and not with respect to the centre of mass of the system.

Adding some 3D to the orbital expressions from earlier in this chapter provides formulas through which the orbit of a double-star system such as α Cen AB can be reconstructed from observables. That may not seem like the most exciting thing to be doing in the 21st century. It turns out, however, that the same principles also apply to some very different systems that started being discovered in the 1990s and are today at the forefront of astrophysics. A common feature of these novel Keplerian systems is that one of the masses is not actually observed, and in fact its very existence is inferred from observing the other mass to be moving on a two-body orbit.

The early discoveries of extrasolar planets were all from measuring the radial velocities of stars and searching for cases where the velocity showed a Keplerian time dependence. (The work of M. Mayor and D. Queloz and of G. Marcy and P. Butler, from 1995 onwards, is well known.) There are also other planet-hunting strategies now, but radial-velocities remain important, especially through extensions of the basic method allowing multiple planets to be discovered, or even planets in multiple-star systems. In particular, the planet α Cen Bb was revealed in this way; as yet there is no image of it, but for an artist's conception see APOD 121018.

Another home of Keplerian orbits is in the very central regions of some galaxies, including the Milky Way. The mass distribution of galaxies is not spherical, and orbits in the gravitational field of a galaxy are generally far from Keplerian, so the discovery of clean Keplerian orbits near galactic centres was a surprise. The inference is that some very concentrated spherical mass lurks at the very centre of a galaxy, making the local gravitational field like $\propto \hat{\mathbf{r}}/r^2$, so compact it can only be a black hole. Near the centre of the Milky Way, a number of stars have been found to be on eccentric Keplerian orbits, in a way reminiscent of periodic comets in the Solar System, but on a much larger scale. A time lapse movie in APOD 001220 shows observations through the 1990s, and animations with more recent data can also be found online.

In distant galaxies, stellar motions such as seen near the Milky Way centre would be too small to measure. But analogous observations can be made in some galaxies, as a result of an extraordinary phenomenon: natural masers. There is always gas present in galaxies, and in a spherical gravitational field, gas tends to settle into discs, with the molecules and atoms themselves moving on nearly circular orbits. Such a configuration is dynamically favoured because it minimizes interatomic collisions and hence reduces dissipation. Gas discs, however, are usually not very bright. In a few cases, notably Messier 106 (a beautiful picture of the galaxy can be seen in APOD 110319, but the central gas disk is much too small to appear on it) there are clouds of water vapour on the disk that act as natural masers. If one then observes with a radio telescope around the maser wavelength of 1.35 cm, one sees bright and comparatively sharply defined clouds, in circular Keplerian orbits.

To understand the maser disk in Messier 106, let us revisit the humble circular orbit, and now include a third dimension. In the orbital coordinate system

$$(x, y, z)_{\text{orbit}} = a\,(\cos\eta, \sin\eta, 0)\,, \qquad \eta = \frac{2\pi}{P_{\text{orb}}}\,t\,, \qquad \frac{d\eta}{dt} = \frac{V_{\text{orb}}}{a}\,. \tag{1.62}$$

To transform to the observing coordinate system, a rotation must be applied. A standard way of expressing the necessary rotation is as a composite of three rotations, as

$$R_z(\omega)\, R_x(I)\, R_z(\Omega)\,. \tag{1.63}$$

Formulas are given in Appendix A, but for the meaning of the sub-rotations, see Fig. 1.3. For a circular orbit, we can drop $R_z(\omega)$ as it is a rotation in the orbital plane, whose only effect is to change the initial phase. The rotation $R_x(I)$ is the important one: I denotes the angle between the angular momentum vector and the line of sight (radial direction), and $R_x(I)$ transforms the orbital plane to an x, y plane tangent to the sky. A further rotation $R_z(\Omega)$ around the line of sight is needed, if (say) we want the x axis to point north. But otherwise, one can drop $R_z(\Omega)$. Applying the rotation matrix $R_x(I)$ gives

$$(x, y, z) = a\,(\cos\eta, \cos I \sin\eta, \sin I \sin\eta) \tag{1.64}$$

and taking dz/dt gives

$$v_z = V_{\text{orb}} \sin I \cos\eta\,. \tag{1.65}$$

The angular sky position will be

$$(\theta_x, \theta_y) = \frac{a}{D}(\cos\eta, \cos I \sin\eta)\,; \tag{1.66}$$

that is, the orbit will appear elliptical on the sky. If (θ_x, θ_y) are observed to change on the sky P_{orb} can be measured, and hence the constants a/D and $\cos I$ can be inferred from (1.66) as well. Then, if v_z is observed to vary with time according to (1.65), P_{orb} and $V_{\text{orb}} \sin I$ can be inferred.

Let us summarize from the above as follows:

$$\begin{aligned} \theta_x(t), \theta_y(t) \quad &\Rightarrow \quad P_{\text{orb}}, \quad I, \quad a/D\,, \\ v_z(t) \quad &\Rightarrow \quad P_{\text{orb}}, \quad V_{\text{orb}} \sin I\,. \end{aligned} \tag{1.67}$$

Let us also recall Eq. (1.8) and recast it as

$$2\pi a = P_{\text{orb}} V_{\text{orb}}\,, \qquad 2\pi G M_{\text{tot}} = P_{\text{orb}} {V_{\text{orb}}}^3\,. \tag{1.68}$$

If all the quantities on the right of Eq. (1.67) are known, then M_{tot} and D can be solved for. Put in another way, if something is on a Keplerian circular orbit, and

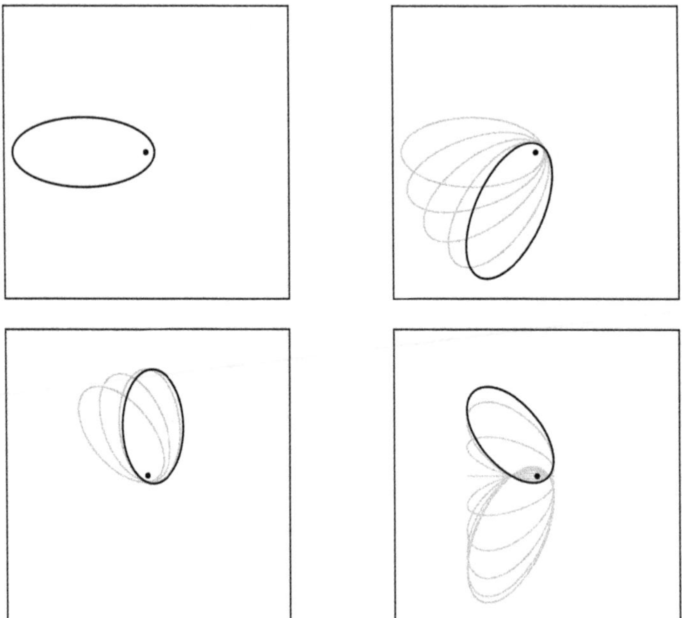

Fig. 1.3 Illustration of the composite rotation Eq. (1.63), using as an example the star S2, which orbits a black hole at the centre of the Milky Way. *Upper left:* the orbit with $a = 6\,\text{light-days}$, $e = 0.88$, $V_{\rm orb} = 0.0058\,c$ in the orbital plane. *Upper right:* rotated by $\omega = 65°$ in the orbital plane. *Lower right:* inclined to the sky by $I = 135°$. (Since $I > 90°$, this rotation changes the apparent motion of the star on the sky from anti-clockwise to clockwise.) *Lower left:* finally, rotated about the line of sight by $\Omega = -40°$, into the standard orientation for sky maps.

the resulting variations in its sky-position and line of sight velocity are measured, the mass of the two-body system and its distance can both be inferred. Messier 106 was the first system for which the distance, and the mass of the central black hole, were measured in this way. Messier 106 itself is ten times the distance of the nearest large galaxy (Andromeda) and maser disk distance measurements now reach ten times further still. We will return to the topic of distances in Chapter 8.

To connect with observables of the stars (mentioned earlier) in orbit around the super-massive black hole at the centre of the Milky Way, we need to consider highly eccentric orbits. More algebra is needed than for circular orbits, but the method remains the same. We start by writing the orbital solution Eq. (1.30) again, with a trivial z dimension:

$$(x, y, z)_{\rm orbit} = a\left(\cos\eta - e, \sqrt{1-e^2}\sin\eta, 0\right) . \tag{1.69}$$

Then we apply the rotation (1.63) as before—see Fig. 1.3 again. This time, $R_z(\omega)$ and $R_x(I)$ are both essential, but $R_z(\Omega)$ can be dropped if the orientation on the sky is not important. For the line of sight velocity

$$v_z = \frac{V_{\text{orb}}}{1 - e\cos\eta}\frac{1}{a}\frac{dz}{d\eta}. \tag{1.70}$$

The black hole in the system is not directly observable itself, as it absorbs all nearby matter and light. Instead, images of the orbiting S stars, with periods ranging from 15 years to several decades, taken over extended durations show these stars tracing out Keplerian orbits, whose measures can be used to estimate the central black hole's mass. Spectra of these stars also enable their velocities along the line of sight to be measured directly.

A further complication arises in the case of α Cen Bb and other extrasolar planets discovered by the same method, which is that the planet itself is not visible. Instead, the planet's existence is inferred from the v_z of its star, which is the other member of the two-body system. Recall from Exercise 1.10 that in a two-body system, the overall orbital system values of a and v are scaled down for each body. Writing $m_\bullet$ as the mass *not* being observed (in this case, the planet), the scaling factor for the observable body's quantities of interest is:

$$\frac{m_\bullet}{M_{\text{tot}}}. \tag{1.71}$$

For planets, typically $m_\bullet/M_{\text{tot}} \ll 1$. For example, in the case of our Solar System, $v \approx 10\,\text{km}\,\text{s}^{-1}$ for Jupiter but comes down to a very small induced velocity for the Sun, $\approx 10\,\text{m}\,\text{s}^{-1}$. The difficulty in viewing such fine effects from far away is the reason extrasolar planets took so long to discover!

Exercise 1.13 Show that the line of sight velocity of one member of a Keplerian two-body system is

$$V_{\text{orb}}\frac{m_\bullet \sin I}{M_{\text{tot}}}\frac{\sqrt{1-e^2}\cos\omega\cos\eta - \sin\omega\sin\eta}{1 - e\cos\eta},$$

where $m_\bullet$ is the mass of the other body, and η is related to time, as usual, through Kepler's equation. Try and plot the line of sight velocity as a function of time, to get a feeling for what the different parameters do—the constant factor on the left sets the overall scale, e decides whether the curve is simply sinusoidal ($e = 0$) or whether it peaks sharply at pericentre (large e), while ω decides how symmetric or asymmetric the curve is. Hence these three constants can be indirectly measured from observations. Since P_{orb} will also be known, show that $(m_\bullet \sin I)^3/M_{\text{tot}}^2$ can also be inferred.

2
Celestial Mechanics

Having solved the two-body problem, Newton turned to a more complicated one: the orbit of the Moon. Now, the Moon is in a nearly Keplerian orbit around the Earth, but since the whole Earth–Moon system itself is in orbit around the Sun, the motions are perturbed by the differences in the Sun's gravitational field that each body experiences. The calculations proved very difficult, and we find Newton complaining in a letter to Halley that the Moon's orbit made his head ache and kept him awake at night... Let us hope that our investigations of such matters here are less pain-inducing.

There are two reasons for Newton's headache. One is that the Moon is comparatively weakly bound to the Earth, so that the perturbations have fairly significant effects. The second and more fundamental reason is that (as noted in Chapter 1) no general, exact solution to the gravitational three-body problem exists. Fortunately for us, though, several special cases *are* soluble. The orbital dynamics of three or more gravitating bodies, studied through approximate treatments or numerical integration, is the topic of celestial mechanics, and it would dominate astrophysics from Newton's time till the early 20th century, and still continues today.

The archetypal celestial-mechanics problem is a special case called the *restricted three-body problem*. Even with limiting 'restrictions', this system already captures the most interesting features of more general gravitational problems. In particular, as Poincaré showed around 1900.[1] (but which only became well-known in the 1960s with the escalating use of computers), the orbit of one of the bodies can be chaotic. In this chapter we will study the restricted three-body problem using general dynamical principles and numerics. Our efforts are rewarded by understanding more about satellite trajectories and dynamics that can turn accretion disks into supernova factories.

2.1 The Restricted Three-Body Problem

The basic arrangement is illustrated in Fig. 2.1. Two bodies, which we will call the primary and secondary, move in circular orbits around their centre of mass, while a third body of negligible mass moves under their combined gravity, in the same plane. The two massive bodies could represent a star and a planet. The third body could be a much smaller planet that orbits the star, or a satellite of the planet, such as a spacecraft.

In order to introduce the orbit equations for the third body, we first introduce some definitions.

[1] For more on this history, see (Diacu and Holmes, 1996).

The Astronomer's Magic Envelope. Prasenjit Saha and Paul A. Taylor.

10.1093/oso/9780198816461.001.0001

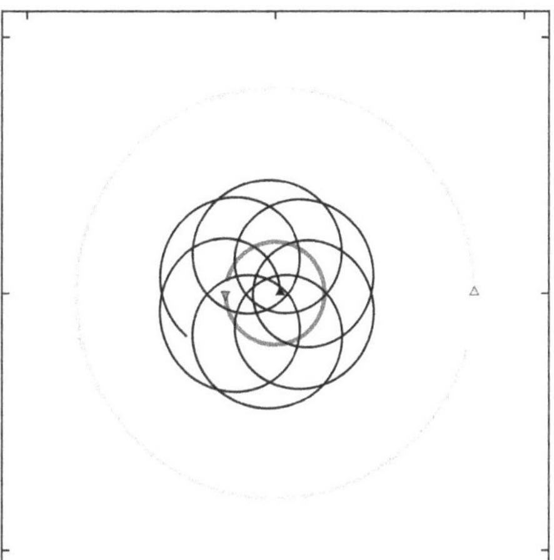
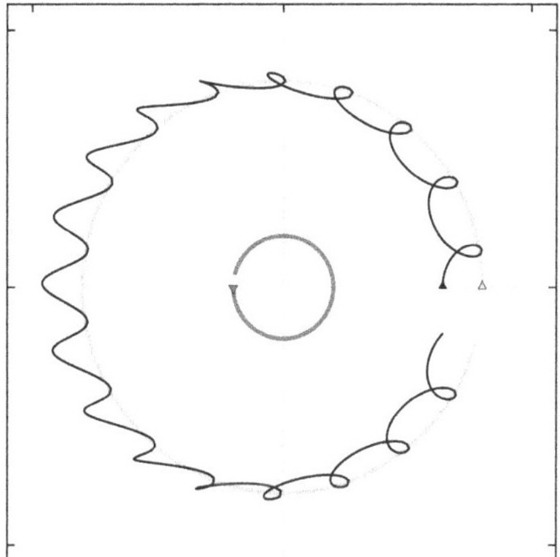

Fig. 2.1 The restricted three-body problem in an inertial (non-rotating) frame. Triangles show the initial locations and directions of the objects. In each case the two massive bodies have a 4:1 mass ratio, with dark and light gray bands showing the paths of the heavier and lighter objects, respectively. The black line shows the path of the third body of negligible mass.

- Let the primary and secondary have masses $(1-\epsilon)M$ and ϵM, respectively, and let their separation be a. Their orbit will be characterized by $P_{\rm orb}$ and $V_{\rm orb}$, as defined in (1.8). It is not necessary for ϵ to be small (only $0 < \epsilon \leq 1$ is assumed) but the regime of $\epsilon \ll 1$ is of particular interest. For example, if the massive bodies were to represent the Sun and Jupiter, then $\epsilon \simeq 10^{-3}$.
- We choose units customized to the system: we let the length unit be a, and the time unit be $\sqrt{a^3/(GM)}$. Any velocity will then come out in units of $\sqrt{GM/a}$, and acceleration in units of GM/a^2.
- We choose a convenient coordinate system. First, the massive bodies are placed at $(-\epsilon, 0)$ and $(1-\epsilon, 0)$, as a consequence of locating the centre of mass at the origin. Second, the two masses *remain* at these values, because we use a rotating reference system.
- As a consequence of the reference-frame choice, our work to obtain the equations of motion of the three bodies in the system is greatly reduced—two of them are stationary by construction, so we only need to derive the trajectory of the massless one.

The gravitational acceleration on the third body due to the masses can now be expressed concisely. Let us write down a potential function, as we did in Section 1.8, of the contributions of the two masses:

$$V(x,y) = -\left(\frac{1-\epsilon}{r_1} + \frac{\epsilon}{r_2}\right), \tag{2.1}$$

where r_1 and r_2 are the distances from the third body to the primary and secondary, respectively:

$$r_1 = \sqrt{(x+\epsilon)^2 + y^2}\,, \qquad r_2 = \sqrt{(x+\epsilon-1)^2 + y^2}\,. \tag{2.2}$$

One can verify that the gradient $-\nabla V(x, y)$ gives the desired acceleration.

With all these definitions, the equations of motion for the third body, in coordinates rotating with the heavy masses, are

$$\begin{aligned} \ddot{x} &= -\frac{\partial V}{\partial x} + x + 2\dot{y}\,, \\ \ddot{y} &= -\frac{\partial V}{\partial y} + y - 2\dot{x}\,. \end{aligned} \tag{2.3}$$

In both lines, the first term comes from the presence of the two masses, and is a 'real force' in the sense that, even if we were using a non-rotating frame, the potential gradient would be present. The second two contributions are sometimes called 'fictitious forces', because they only arise when we choose to work within the rotating coordinate grid. The first of these is the centrifugal acceleration, which is position-dependent and acts radially outwards (i.e., it appears to 'push' third body away from the centre in this frame of reference). The second ficticious term is the Coriolis acceleration, which is velocity-dependent and torques the particle's trajectory azimuthally (i.e., pushes it angularly within the orbital plane) whenever it has any radial motion.

Exercise 2.1 How would the equations of motion Eq. (2.3) have appeared, had we not chosen custom units to eliminate GM and a? (One simply needs to put in appropriate powers of length and time units to make the dimensions consistent.)

Exercise 2.2 A planet, even though it is much less massive than its star, will dominate dynamics close to it. This is the regime that satellites often inhabit. To study it, we shift the origin by introducing $X = x - (1-\epsilon)$, and then put ϵ, X, $y \ll 1$. Show that the equations of motion (2.3) reduce to

$$\begin{aligned} \ddot{X} &= -\epsilon\frac{X}{R^3} + 2\dot{y} + 3X\,, \\ \ddot{y} &= -\epsilon\frac{y}{R^3} - 2\dot{X}\,, \end{aligned}$$

where $R = (X^2 + y^2)^{1/2}$. These are known as *Hill's equations.* A related system appears in spacecraft dynamics, where it is called the Clohessy-Wiltshire model.
Hint: The expansion $(1+\alpha)^{-1/2} = 1 - \frac{1}{2}\alpha + \frac{3}{8}\alpha^2 + \mathcal{O}(\alpha^3)$ is useful here.

2.2 Lagrange Points, Roche Lobes and Chaotic Systems

In Fig. 2.2 we revisit the setup and orbits from Fig. 2.1 in a rotating frame. Figure 2.2 also illustrates a new effective potential that can be constructed, called the Roche potential

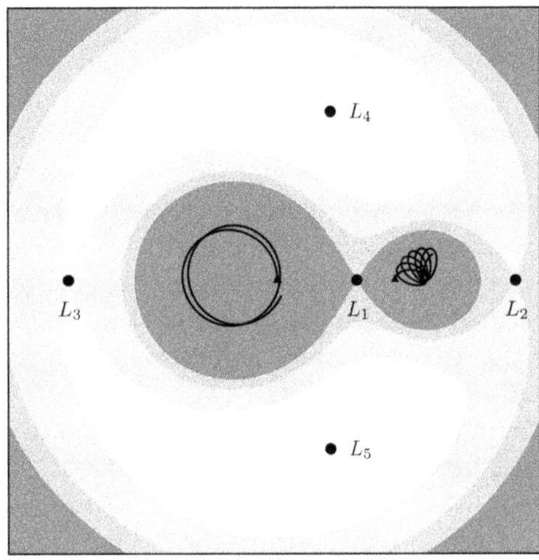

Fig. 2.2 The trajectories from Fig. 2.1 (but run only a third as long) transformed to a frame rotating with the two massive bodies. The boundaries of the different shaded regions are contours of the Roche potential V_R in Eq. (2.4). The two touching lobes are the Roche lobes. Filled circles mark the Lagrange points, which are zero-gradient points of the Roche potential.

$$V_{\rm R}(x, y) = -\left(\frac{1-\epsilon}{r_1} + \frac{\epsilon}{r_2}\right) - \tfrac{1}{2}(x^2 + y^2)\,. \tag{2.4}$$

This describes the combined effects of the gravitational (first term) and centrifugal (second term) accelerations, but it does not include the Coriolis acceleration. The Roche potential has several interesting consequences, which qualitatively can be read from a plot of it (see Fig. 2.2), looking for maxima and points with any stability.

First, because the gravitational force is directed inward from each of the two massive bodies, while the centrifugal force is always outward, we might expect there to be locations where the forces cancel in at least one direction. There are in fact five such points, known as Lagrange points, and they are individually named as L_1 through L_5. If the third body is *exactly* at a Lagrange point and not moving (with respect to that frame), then it will remain there. If it is merely close to a Lagrange point, it may or may not remain so. We explore this question of stability in more detail later.

Second, the effect of the Roche potential is not limited to an isolated orbiting body. If the primary or secondary has a fluid around it, such as an atmosphere or accretion disk, then that fluid will tend to be shaped according to the contours of the Roche potential. Forces are normal to potential contours, and therefore Fig. 2.2 indicates that the fluid will tend to form oval lobes (the dark-shaded regions), with the fluid surfaces of constant density normal to the potential in equilibrium. Even on Earth, this influence is observable: the oceans feel the influence of the Roche potential, which is why tides form on the Earth.

The largest lobes of the potential, where the atmospheres of the primary and secondary would touch, are known as Roche lobes. These are important in binary-star systems (and it should be noted that perhaps half of all stars are in binaries). In close binary stars, one star can overflow its Roche lobe, resulting in some material being donated to the other star. This mass transfer alters the size of the lobes themselves, making the donor's Roche lobe smaller and the recipient's larger. The relative change in mass of the objects shifts the potential, in some cases in a dynamically unstable way: it drives further mass transfer, which shifts the potential more, feeding more mass transfer, etc. A runaway mass transfer may result, and in some scenarios the sudden influx of mass can destabilize the recipient star, leading it to self-destruct. This phenomenon may be the driver of many supernovae, such as the historical supernova SN 185 (APOD 060928 and APOD 111110).

Let us now consider the Lagrange points in more detail. To locate them, we take the acceleration implied by the Roche potential

$$\begin{aligned} -\frac{\partial V_{\mathrm{R}}}{\partial x} &= x - \frac{(1-\epsilon)(x+\epsilon)}{(r_1)^3} - \frac{\epsilon(x+\epsilon-1)}{(r_2)^3}\,, \\ -\frac{\partial V_{\mathrm{R}}}{\partial y} &= y\left(1 - \frac{1-\epsilon}{(r_1)^3} - \frac{\epsilon}{(r_2)^3}\right), \end{aligned} \tag{2.5}$$

and search for coordinates where both acceleration components are zero. Putting $y = 0$ in (2.5) makes the y-acceleration zero. It also simplifies r_1 and r_2 to $|x+\epsilon|$ and $|x+\epsilon-1|$, respectively. The condition for the x-acceleration to be zero is then

$$x \pm \frac{1-\epsilon}{(x+\epsilon)^2} \pm \frac{\epsilon}{(x+\epsilon-1)^2} = 0\,. \tag{2.6}$$

There are three possibilities for the two '$\pm$' signs, depending on the particle's location relative to the large masses, and these give three Lagrange points. If $x < -\epsilon$, both primary and secondary are pulling to the right; hence, both signs are positive. The point is known as L_3. If $-\epsilon < x < 1-\epsilon$, the primary term becomes a negative (L_1). If $x > 1-\epsilon$, both the $\pm$ are negative (L_2). In each case, the equation is a quintic in x and hence can only be solved numerically. But if ϵ is small, we can apply Hill's approximation of Exercise 2.2 to find that

$$x \approx 1 - \epsilon \pm (\epsilon/3)^{1/3} \tag{2.7}$$

for two Lagrange points near the secondary. The distance $(\epsilon/3)^{1/3}$ is called the *Hill radius.*

From (2.5), we might also note that any point satisfying $r_1 = r_2 = 1$ will also have vanishing gradients. Hence the points which happen to make an equilateral triangle with the primary and secondary will also be Lagrange points. These points are $(x, y) = (-\epsilon + 1/2, \pm\sqrt{3}/2)$, with L_4 leading and L_5 trailing the large masses.

Near the axial Lagrange points (L_1, L_2, and L_3), the $\ddot{y}$ of the Roche potential is oriented towards the Lagrange point, while its $\ddot{x}$ pushes away. In other words, these locations are saddle points of the effective potential and are unstable: a small displacement from L_i along the x-axis tends to get pushed successively further away. Moreover, near a saddle point, trajectories tend to become more sensitive to initial conditions: points starting off arbitrarily close can end up in very different locations. As a result, the natural dynamics around these points are chaotic. Orbits that go near one of the axial Lagrange points tend to become indecisive about where they want to go. That may be bad news from some perspectives, but it can also be useful if the third body is not purely passive. Imagine that a spacecraft is very near one of these unstable Lagrange points, and that it is ready to wander off in some particular direction. Using a very small but carefully calculated thrust, it can just nudge itself to a particular starting point, and then the effective potential will guide it along its desired path without any more fuel being spent. Alternatively, a spacecraft can maintain a position near an unstable Lagrangian point for a long time by spending small amounts of fuel. This strategy is used by some space missions wanting to observe the sky without the Earth filling too much of the field of view. The Sun–Earth L_1 is ideal for observing the Sun from a safe distance, and is the location of the SOHO spacecraft APOD 051201. Space observatories keen to avoid the infrared blaze from both the Sun and the Earth seek out the Sun–Earth L_2: thus, Herschel APOD 160114, Planck APOD 100322, and the upcoming James Webb Space Telescope APOD 170318.

The remaining Lagrange points L_4 and L_5 sit at maxima of the effective potential. Being maxima, one would expect them to be absolutely unstable, but in fact, their neighbourhoods can harbour stable orbits, provided ϵ is small enough. The reason for this is that the Coriolis acceleration, which recall was not included in the effective potential, plays a stabilizing role. To investigate, let us introduce coordinates (X, Y) with respect to L_4 and L_5:

$$x = -\epsilon + \frac{1}{2} + X, \qquad y = \pm\frac{\sqrt{3}}{2} + Y\,. \tag{2.8}$$

Here the plus of the $\pm$ refers to L_4 and the minus refers to L_5. Returning to the full equations of motion (2.3) and expanding them to first order in X, Y, ϵ (computer algebra helps here), we obtain the following messy but linear system:

$$\begin{aligned} \ddot{X} &= \frac{3}{4}X \pm \frac{3\sqrt{3}}{4}(1-2\epsilon)Y + 2\dot{Y} \\ \ddot{Y} &= \frac{9}{4}Y \pm \frac{3\sqrt{3}}{4}(1-2\epsilon)X - 2\dot{X}\,. \end{aligned} \tag{2.9}$$

Being linear, the differential equations (2.9) can be analysed for normal modes of the form $X = X_0 \exp(\gamma t), Y = Y_0 \exp(\gamma t)$. We will not go into the labour-intensive details here, but the result is that if ϵ is small and specifically $\epsilon(1-\epsilon) < 1/27$, then γ must be pure imaginary, implying that the normal modes are oscillatory and not growing. Hence L_4 and L_5, despite being maxima of the Roche potential, are stable. Several small objects in the Solar System are known to hang out in the vicinity of L_4 or L_5 with respect to two much more massive bodies. Archetypical are the Trojan asteroids, which are associated with the Sun–Jupiter system. The Earth has a few associated Trojans too. An even more exotic example is the Saturnian moon Helene (APOD 130105), which orbits near L_4 of Saturn and the moon Dione.

Exercise 2.3 Choosing some ϵ, compute L_1, L_2, and L_3. Plot the contours of the effective potential that pass through these points. For $\epsilon = 0.2$ you can compare your results to the grey shades in Fig. 2.2.

2.3 Hamilton's Equations and Interesting Orbits

Celestial mechanics, and the restricted three-body problem in particular, are often studied using the language of Hamiltonians.[2] Hamiltonian dynamics is often presented as rich in mathematical complexity and low in usefulness, but in fact is quite convenient for applications such as in this book. At the heart of Hamiltonian formulations are two things. The first is that for each coordinate, a momentum variable is introduced, which has a complementary (the technical term is 'conjugate') role. In our case, x, y are joined by the momentum variables p_x, p_y. The two second-order equations (2.3) for $(\ddot{x}, \ddot{y})$ are replaced by four first-order equations for $(\dot{x}, \dot{y}, \dot{p}_x, \dot{p}_y)$. The second thing is that the equations of motion are encoded as a single scalar function of the position and momentum variables: the Hamiltonian H. Naturally, not every system of differential equations can be cast in Hamiltonian form, but many physically important dynamical systems can.

Hamiltonian formulations bring three advantages. First, it is possible to identify conserved quantities directly. Second, transforming equations of motion between coordinate systems—say between Cartesian and polar, or between inertial and rotating—proceeds simply and elegantly. Third, the equations of motion themselves, being coupled first-order equations, take a form well-suited to numerical solution. For these reasons, it is interesting and useful to see the restricted three-body problem in Hamiltonian garb. In the next chapter, Hamiltonians will become even more useful.

With that preamble, here is the Hamiltonian equivalent to our earlier equations of motion in Eq. (2.3) for the restricted three-body problem:

$$H(x, y, p_x, p_y) = \tfrac{1}{2}\left(p_x^2 + p_y^2\right) + V(x, y) + (y p_x - x p_y). \tag{2.10}$$

The Hamiltonian has a kinetic and potential part, as in the example in Eq. (B.5) from Appendix B, followed by terms arising from a rotating frame, as explained in the last part of that appendix. The Hamiltonian equations of motion are generated by applying Hamilton's equations from Eqs. (B.4), which give

$$\begin{aligned} \dot{x} &= p_x + y, & \dot{p}_x &= p_y - \frac{\partial V}{\partial x}, \\ \dot{y} &= p_y - x, & \dot{p}_y &= -p_x - \frac{\partial V}{\partial y}. \end{aligned} \tag{2.11}$$

We see here that (p_x, p_y) differ from the kinematic momentum $(\dot{x}, \dot{y})$. The (p_x, p_y), which are known in Hamiltonian dynamics as the canonical momentum variables, somehow know that we are not in an inertial frame and hence things are different.

[2] Appendix B gives a very short introduction.

Another interesting consequence from Hamiltonian dynamics is that because this $H(x, y, p_x, p_y)$ has no explicit functional dependence on time, it is constant along any orbit. Because we are in a rotating frame, potential plus kinetic energy in the frame are not constant, but H is. It is known as the Jacobi constant or Jacobi integral.

Numerical integration of Eqs. (2.11) reveals some fascinating orbits, some of which you can explore numerically from Exercise 2.5 at the end of this chapter, and of which real-life examples continue to be discovered in the Solar System. We have already mentioned the Trojans, which in an inertial frame are almost on the same orbit as the secondary, but 60° behind or ahead in orbital phase. In rotating coordinates, Trojans move on out a curly trajectory whose envelope roughly traces a contour of the effective potential near L_4 or L_5.

Additionally, there are orbits whose period is in some simple ratio (such as a half) with the secondary. These are called resonant orbits. One surprising category of resonant orbits are retrograde co-orbitals: these have the same orbital period as the secondary, but go around the other way. In a frame rotating with the secondary, retrograde co-orbitals trace a loop-within-loop pattern (known as a limaçon) and avoid colliding with the secondary. Before the 21st century, not even contrarian theorists imagined asteroids orbiting backwards, but in fact there is (at least) one, so far known only as 2015 BZ_{509}.

Another exotic kind of orbit are horseshoes, so named because they move in a kind of horseshoe pattern enveloping L_4, L_3, L_5. Horseshoes are not retrograde, meaning that in inertial coordinates they go around the same way as the secondary. But in the rotating frame they go backwards as well as forwards. There are a few asteroids known in horseshoe orbits with respect to the Sun–Earth, but the best-known examples are the Saturnian moons Janus and Epimetheus (see APOD `051102`). Because the two moons have comparable masses, that system does not quite correspond to the restricted three-body problem; rather, Janus and Epimetheus share the roles of secondary and third body in a three-body system with Saturn.

Exercise 2.4 From the Hamiltonian equations (2.11), derive the acceleration equations (2.3).

Exercise 2.5 Integrate Eqs. (2.11) for different choices of ϵ and initial conditions. Some interesting examples to try are

ϵ	x	p_y
0.2	0.0	2.0
0.2	0.64	1.20068
0.01	0.334134	2.032871
0.01	−0.853236	−0.575821
0.01	0.023867	7.472068
0.001	1.9	−0.23
0.000953875	−1.06206	−0.96941

all with $y = p_x = 0$ initially. The first two give the examples shown in Fig. 2.2. The next three give curious-looking but rather beautiful patterns with two, three and four leaves. The one after is a retrograde co-orbital; this example is from a paper by (Morais and Namouni, 2016). The last one, perhaps the most striking of all, is a horseshoe orbit; it is taken from a paper (Taylor, 1981).

In addition, you can explore what happens near the various Lagrange points. To do so, place the third body at a Lagrange point with a very small velocity. Hamilton's equations (2.11) tell us that small velocity does not mean small momentum, but rather (p_x, p_y) nearly $(-y, x)$.

3
Schwarzschild's Spacetime

Throughout the 18th and 19th centuries, the gradual accumulation of telescopic observations led to ever more precise measurements of orbits within the Solar System. Meanwhile, celestial mechanists developed methods for calculating perturbations of orbits by other planets. By 1860, a tiny discrepancy was evident. Theory predicted that the eccentricity vector of Mercury should precess by 2π about once every 245,000 years, because of perturbations by the other known planets (mainly Jupiter). But the observed precession rate was about 10% faster. Was there an undiscovered planet, further in than Mercury, providing the extra perturbation? A mysterious planet called Vulcan acquired several claimed 'discoveries' over a few decades but ultimately lived on only in science folklore and fiction.

Instead, developments in other areas of physics implied that a new theory of gravity was needed, of which Newtonian gravity would be a limiting case. This movement culminated with Einstein providing such a vehicle in 1915—the theory of general relativity. Seemingly miraculously, it provided just the right amount of precession[1] needed to explain Mercury's orbit—the first of many successes.

J. A. Wheeler has summarized general relativity thus:[2]

Spacetime tells matter how to move;
matter tells spacetime how to curve.

This chapter is devoted to an example of the first part of Wheeler's epigram, dynamics in a relativistic framework. We will explain what spacetime is and what it means for space and spacetime to be curved. We will then see the general relativistic version of Newton's apple problem. This much can be done using the formalism of Hamiltonian dynamics, which we have already used for celestial mechanics. The second part of the Wheeler quotation requires the full apparatus of general relativity, which we cannot hope to cover here,[3] so we will just take over one key result: a formula describing space and time around a spherical mass, derived by Karl Schwarzschild in 1916.

[1]The figure of 43 arc-seconds per century has by now passed into popular culture.

[2]From his autobiography (Wheeler, 1998). Wheeler also popularized the term 'black hole'.

[3]The traditional style of textbooks on general relativity, and of comparable online resources, has been to develop all the mathematical formalism before touching any astrophysical applications. Some recent books, however, develop the formalism more in step with the physics. For a guide to the landscape of general-relativity texts, see the article by (Christensen and Moore, 2012).

The Astronomer's Magic Envelope. Prasenjit Saha and Paul A. Taylor.

10.1093/oso/9780198816461.001.0001

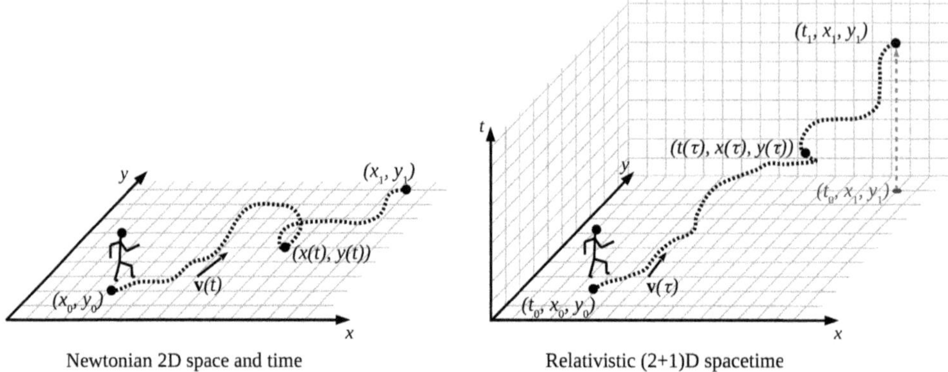

Fig. 3.1 Motion in the Newtonian (*left*) and relativistic (*right*) points of view. In the Newtonian case, the time t is a parameter along the path through the spatial coordinates. In relativity, time *is* also a coordinate itself, and new variable, the proper time τ, parametrises the path.

3.1 Spacetime and Proper Time

A fundamental entity within relativity theory is four-dimensional *spacetime*, the usual three spatial dimensions with time included as the fourth. Let us elaborate upon what this means, with the help of the cartoon in Fig. 3.1.[4]

In the left half of the figure, we have a Newtonian athlete running from (x_0, y_0) to (x_1, y_1). Clocks measuring time t are present all over the landscape. A clock carried by the athlete always agrees with the local time, and the athlete's velocity is unambiguously defined as $\mathbf{v}(t) = (dx/dt, dy/dt)$.

Nature, however, does not quite follow Newtonian dynamics: time does not elapse the same everywhere. If the clocks are accurate enough, our athlete moving at speed $v = |\mathbf{v}| > 0$ will notice that their own clock seems to be slower than the landscape clocks. The effect is of order v^2/c^2, where c is the speed of light. Thus, at sprinting speeds, the discrepancy will be of order 10^{-16}, which these days is just at the threshold of measurability. It should be noted that this discrepancy between the clocks is not some mechanical effect: *any* time-measuring process shows the same effect. Time really slows down.

In relativity theory the way around the problem is to make time an extra dimension, as illustrated on the right in Fig. 3.1. Each point in space has an independent clock measuring t. The athlete runs from an *event* (t_0, x_0, y_0) to another event (t_1, x_1, y_1). A clock carried by the athlete measures the so-called *proper time* τ , which is something that all clocks can agree on. The path is described by $(t(\tau), x(\tau), y(\tau))$.

The cartoon example and picture suggests that relativistic motion can be expressed using the same mathematical apparatus as classical dynamics, if we just identify quantities correctly: t gets promoted to coordinate status, and τ plays the role of the parameter. Indeed, as we will see later, we can apply this relativistic framing to orbit

[4]An additional example and discussion are provided in Appendix C.

equations and then write them in Hamiltonian form. As part of t being identified as a coordinate, it will also acquire its own conjugate momentum p_t.

For the rest of this chapter (and only within this chapter), we will use dots to denote derivatives with respect to τ and not with respect to t:

$$(\dot{t},\, \dot{x},\, \dot{y},\, \dot{z}) \equiv \left(\frac{dt}{d\tau}, \frac{dx}{d\tau}, \frac{dy}{d\tau}, \frac{dz}{d\tau}\right). \tag{3.1}$$

3.2 Metrics

Let us recall for a moment one of the most fundamental of all principles in classical dynamics: Newton's first law. One way of stating the law is that a body with no external force acting on it takes the shortest path between two points. In the language of spacetime and events, we can say the following: an isolated body takes a path Δs between two events such that

$$\Delta s^2 = -c^2 \Delta t^2 + \Delta x^2 + \Delta y^2 + \Delta z^2 \tag{3.2}$$

is extremal.[5] (The sign of Δs^2 is not important here: $\Delta s = \sqrt{|\Delta s^2|}$.) We have in (3.2) a new and curious definition of distance, known as the Minkowski distance (after Hermann Minkowski, an older contemporary of Einstein). Einstein made the inspired guess that motion in gravitational fields could be understood by retaining the idea of extremal distances between events, while generalizing the Minkowski distance by introducing curvature.

The essential notions are in fact already familiar to us. Imagine a long journey on the surface of the Earth: we travel straight ahead for 10,000 km, then turn right and travel forward another 10,000 km, then turn right again and travel yet another 10,000 km forward, and find that we have returned to our starting point. Throughout this journey, the Earth would have seemed flat every step of the way (what is more formally called being *locally flat*). Yet, the trip as a whole would tell us that the Earth's surface is curved: we took three 90° turns after travelling equal distances, and we came to the beginning.

That example had only a single reference point: the starting (and ending) point on the Earth. Suppose now, that we had many reference points, say all the towns and villages over a continent—an arbitrary coordinate system. By measuring the distances between the reference points, again assuming *local* flatness, we could measure the *global* curvature of the Earth.

More formally, on a unit sphere with polar coordinates on it, the distance between two points $(\theta_0,\, \phi_0)$ and $(\theta_1,\, \phi_1)$ is

$$\int_{\theta_0,\, \phi_0}^{\theta_1,\, \phi_1} ds\,, \tag{3.3}$$

where the differential distance ds is given by

[5] We use the following common abbreviation here and in Appendix C: $\Delta s^2 \equiv (\Delta s)^2$, $ds^2 \equiv (ds)^2$, etc. This greatly improves the readability of metric expressions.

$$ds^2 = d\theta^2 + \sin^2\theta \, d\phi^2 \,. \tag{3.4}$$

An expression of the type (3.4) is known as a *metric equation.* As we argued earlier, curvature is implicit in the metric—the two concepts are inseparable. The curvature information does not depend on being embedded in higher dimensions. Thus, a football that has been deflated and folded still retains its intrinsic curvature, and a crumpled sheet of paper is geometrically still a plane.

Now let us consider an example of a three-dimensional curved space. The coordinates are r, θ, ϕ, with $r > 1$, and the metric equation in terms of these is

$$ds^2 = \frac{dr^2}{1 - 1/r} + r^2 \left(d\theta^2 + \sin^2\theta \, d\phi^2\right) \cdot \tag{3.5}$$

Surfaces of constant r are spherical, as we already know. But there is also curvature along r. Consider two circles at $\theta = \pi/2$ (equatorial), one having $r = r_1$ and the other at $r = r_2 > r_1$. The minimum distance between these circles is

$$\int_{r_1}^{r_2} \frac{dr}{\sqrt{1 - 1/r}} \tag{3.6}$$

which is more than $r_2 - r_1$. Space seems to get somehow more spacious as r decreases, eventually becoming singular at $r = 1$. Space near a black hole has such a property.

In general relativity the *metric* $g_{\alpha\beta}$ is the quantity that defines the spacetime, and all its properties are derived from it. It contains the information of how the system's coordinates combine to form the line element, some examples of which we have seen earlier. In general form, the metric equation, starring $g_{\alpha\beta}$ in the central role, is

$$ds^2 = \sum_{\alpha,\, \beta} g_{\alpha\beta} \, dx^\alpha \, dx^\beta \,. \tag{3.7}$$

Here, x^α stands for any of the four spacetime coordinates, such as any of $\{t,\, x,\, y,\, z\}$ in Cartesian coordinates, for which then $x^0 \equiv t, x^2 \equiv y$, etc. Each element of the 4×4 matrix $g_{\alpha\beta}$ is a function of those coordinates.

Let us look again at the example of Minkowski distance and spacetime from above in Eq. (3.2) with this new information in mind. The metric equation in this case could be expressed in either Cartesian or spherical polar (or other) coordinates, respectively:

$$\begin{aligned} ds^2 &= -c^2\, dt^2 + dx^2 + dy^2 + dz^2 \\ &= -c^2\, dt^2 + dr^2 + r^2 d\theta^2 + r^2 \sin^2\theta \, d\phi^2 \,. \end{aligned} \tag{3.8}$$

The metric itself can be read from these equations and written for the first and second lines, respectively, as

$$g_{\alpha\beta} = \begin{pmatrix} -c^2 & 0 & 0 & 0 \\ 0 & 1 & 0 & 0 \\ 0 & 0 & 1 & 0 \\ 0 & 0 & 0 & 1 \end{pmatrix} \quad \text{or} \quad g_{\alpha\beta} = \begin{pmatrix} -c^2 & 0 & 0 & 0 \\ 0 & 1 & 0 & 0 \\ 0 & 0 & r^2 & 0 \\ 0 & 0 & 0 & r^2 \sin^2\theta \end{pmatrix} \cdot \tag{3.9}$$

There is no curvature here—the metric is flat because we can write it in *a* form (here, the Cartesian one) in which every element is constant (derivatives with respect to the system coordinates are zero); this property holds in whatever set of coordinates we write it. The variable terms in the second (spherical polar) form can be used to describe curved subspaces associated with θ and ϕ, but they do not imply curvature as a whole. (A complicated coordinate system can result in a complicated-looking metric, but curvature does not depend on the choice of coordinate system. The choice of coordinates can't change the geometric properties of the space, but it can make them more apparent. As with selecting units, we pick coordinates for convenience in certain circumstances.)

Einstein's field equations are differential equations for $g_{\alpha\beta}$ that provide the formal description of how 'matter tells spacetime how to curve'. Only a few exact solutions of Einstein's field equations are known. The earliest and best-known of these is Schwarzschild's solution:

$$\begin{aligned} ds^2 = &-\left(1-\frac{2GM}{c^2 r}\right) c^2\,dt^2 + \left(1-\frac{2GM}{c^2 r}\right)^{-1} dr^2 \\ &+ r^2\left(d\theta^2 + \sin^2\theta\, d\phi^2\right), \end{aligned} \tag{3.10}$$

which describes the spacetime around a point mass M at the origin. In fact, this description is even valid in some cases when M is not a point mass (similarly to the Newtonian case discussed in Chapter 1): any isolated, spherical, non-rotating mass will have Schwarzschild spacetime described by (3.10) outside it.

The path of a test particle in spacetime with metric $g_{\alpha\beta}$ is such that $\int ds$ is a minimum, or at least extremal. There are several ways of finding an extremal path. One elegant and very general prescription is as follows. Construct the function

$$H = \tfrac{1}{2}\sum_{\alpha,\beta} g^{\alpha\beta}\, p_\alpha\, p_\beta\,, \tag{3.11}$$

where $g^{\alpha\beta}$ is the inverse matrix of $g_{\alpha\beta}$. If this H—which so far is just a specific invented function of the coordinates x^α and some new variables p_α—is interpreted as a Hamiltonian with τ as the independent variable, then the trajectories give the extremal paths of $\int ds$. H has no explicit dependence on τ and therefore must be constant along a particle trajectory. Trajectories with $H < 0$ apply to ordinary matter, and those with $H = 0$ apply to light. The case of $H > 0$ would imply travel faster than light, which is not mathematically *forbidden* by relativity, but generally considered unphysical.

Exercise 3.1 In cylindrical coordinates $(r,\ \phi,\ z)$, the surface

$$r = e^z, \qquad 0 \le z \le 1,$$

is roughly a vuvuzela shape, for which one can adopt (r, ϕ) as a coordinate system. Show that

$$ds^2 = (1 + 1/r^2)\, dr^2 + r^2\, d\phi^2.$$

As in a Schwarzschild metric, the distance between two circles at $r = r_1, r_2$ on a vuvuzela is not $r_2 - r_1$.

3.3 Schwarzschild Orbits, I: General Properties

The two preceding sections summarize some important results from general relativity: matter in the form of a spherical, non-rotating mass M tells spacetime to curve in the way described by the Schwarzschild metric (3.10); this spacetime in the Schwarzschild shape prescribes a Hamiltonian for the motion of particles (as any spacetime/metric would). We have not derived these results *ab initio* from ground principles, having instead merely taken them over from books on general relativity. But having done that, we will now deduce some consequences of the theory that are relevant.

So, we now want to investigate particles moving around a non-rotating, spherical mass M (i.e., Schwarzschild's spacetime). As in the previous chapter, we will use special units to make the equations simpler. In this case we will choose a scale in which the magnitude of the speed of light is unity, so that $c = 1$, and the strength of gravity is such that $G = 1$ (these selections foreshadow a particular set of units introduced in Chapter 4 and widely used thereafter). As a consequence, it will appear that we are measuring time in length units—in other words, when we write t, we really mean ct. Additionally, and perhaps surprisingly, mass will also look like it has units of length, as well—that is, when we write M, we really mean GM/c^2. Then, the following Hamiltonian is generated by taking the spherical polar coordinate form of the metric in Eq. (3.9) and applying the formulation from Eq. (3.11):

$$\begin{aligned} H = &-\left(1-\frac{2M}{r}\right)^{-1}\frac{p_t^2}{2}+\left(1-\frac{2M}{r}\right)\frac{p_r^2}{2} \\ &+\left(\frac{p_\theta^2}{2r^2}+\frac{p_\phi^2}{2r^2\sin^2\theta}\right). \end{aligned} \tag{3.12}$$

Since this Hamiltonian has no explicit dependence on t, then

$$\dot{p}_t = -\frac{\partial H}{\partial t} = 0\,, \tag{3.13}$$

and hence p_t must be a constant. The value of p_t is not physically important—in effect, it just sets the units for τ—so we choose to simply let $p_t = -1$ (again, selecting units to make notation as uncluttered as possible). By Eq. (B.4), this gives

$$\dot{t} = (1-2M/r)^{-1} \tag{3.14}$$

(again, recalling that in this chapter dotted variables have their derivative taken with respect to τ). This means that coordinate time t, compared to proper time, goes faster as we approach the mass from far away (verify this by comparing some $\dot{t}$ values when $r > 2M$). That is to say, a clock in a near orbit runs more slowly than one in a distant orbit, or proper time is relatively slower deep in the gravitational field. For artificial satellites around the Earth, the effect is $\sim 10^{-9}$, which is easily measurable with modern instruments such as the clocks on GPS satellites.

We now verify that the Schwarzschild Hamiltonian has the correct Newtonian limit. Such nonrelativistic particles by definition must be moving slowly compared to the speed of light ($p_r \ll 1$, in our present units) and be far away from the centre ($r \gg M$).

Accordingly, let us consider the leading-order contributions involving p_r^2 and M/r. Expanding out the factor $(1 - 2M/r)^{-1}$ and using our unit-based choice for $\dot{p}_t$ from earlier leads to

$$H = -\frac{1}{2} - \frac{M}{r} + \frac{p_r^2}{2} + \frac{p_\theta^2}{2r^2} + \frac{p_\phi^2}{2r^2 \sin^2\theta} + \mathcal{O}\left(\frac{M^2}{r^2} + \frac{M}{r}p_r^2\right). \tag{3.15}$$

This, indeed, is the spherical polar-coordinate Hamiltonian for a classical particle in a potential $V(r) = -M/r$, plus higher-order terms—*and* the additional constant $-\frac{1}{2}$. The constant drops out in Hamilton's equations, which are all based on derivatives, so it has no dynamical effect. But it reminds us that $H \to -\frac{1}{2}$ for a static particle at $r \to \infty$—in contrast, recall that $H = 0$ corresponds to light. The numerical value of H itself is not the orbital energy, and instead $H + \frac{1}{2}$ plays that role.

Let us look at orbital motion around the Schwarzschild mass for the fully relativistic case whose Hamiltonian is given by Eq. (3.12) (also keeping our earlier unit choice, $p_t = -1$). Let us start with motion in the equatorial plane of the coordinate system, and see what happens with the dynamics. That is, at some instant of (proper) time, we choose the initial conditions: polar angle $\theta = \pi/2$ (so that we are the equatorial plane) and $\dot{\theta} = 0$ (so that we are initially moving *within* the plane). What happens with the polar angle variables? Well, from the Hamilton definitions in Eqs. (B.4), we have

$$\dot{\theta} = \frac{\partial H}{\partial p_\theta} = \frac{p_\theta}{r^2}, \qquad \dot{p}_\theta = -\frac{\partial H}{\partial \theta} = \frac{\cos\theta}{\sin^3\theta}\frac{p_\phi^2}{r^2}. \tag{3.16}$$

With our initial condition, it follows from the first equation that $p_\theta = 0$ and from the second that $\dot{p}_\theta = 0$. That is, a particle moving within the equatorial plane *stays* confined within it. As a result we can discard the p_θ term in the Hamiltonian without loss of generality. (NB: we can always rotate coordinates to locate motion within a defined equatorial plane in order to simplify things, without altering any meaning; any orbital plane *remains* the orbital plane, and we can choose our coordinates to simplify our work.) This leaves the simplified Hamiltonian expression as:

$$H(r, \phi, p_r, p_\phi) = \left(1 - \frac{2M}{r}\right)\frac{p_r^2}{2} + \frac{p_\phi^2}{2r^2} - \frac{r/2}{(r - 2M)}. \tag{3.17}$$

We note again that $H = 0$ corresponds to light, whereas ordinary bodies have $-\frac{1}{2} < H < 0$, and $H + \frac{1}{2}$ is the orbital energy.[6]

We have so far ignored the fact that something very strange happens for this system at $r = 2M$ (called the *Schwarzschild radius*). Consider the expression in Eq. (3.14), which we could rewrite as

$$\dot{t} = \frac{r}{r - 2M}. \tag{3.18}$$

In this form it is apparent that the coordinate time runs 'infinitely slow' as one approaches there: for each small change in proper time, there is an increasingly large interval of t, implying that it would take an infinite amount of time to reach $r = 2M$

[6] Actually *specific* orbital energy, or orbital energy per unit mass. In this chapter energy, momentum, and angular momentum are always per unit mass of the small orbiting body.

or to get anywhere from there. In fact, one *can* get there in finite proper time, but no *information* can come out in finite time. This property leads to the phenomenon of a black hole being 'black': when the mass M of an object is entirely contained within its Schwarzschild radius, then no mass-carrying particle or photon of light can be emitted. That is, events which take place within $r < M$ are unknown to the outside Universe; the Schwarzschild radius is what defines the edge of a black hole, or the *event horizon.*

Exercise 3.2 Using the information that the Earth–Sun distance $\simeq$ 500 light-sec and the Earth's orbital period is a year, compute the Schwarzschild radius for a solar mass. Convert it to km.

3.4 Schwarzschild Orbits, II: Circles

The Hamiltonian in (3.17) has no explicit dependence on the azimuthal angle ϕ, from which it follows (again, from Eqs. (B.4)) that p_ϕ is a constant. Hamilton's equations for the remaining (radial) quantities are

$$\begin{aligned} \dot{p}_r &= \frac{p_\phi^2}{r^3} - \frac{M}{(r-2M)^2} - M\frac{p_r^2}{r^2}, \\ \dot{r} &= p_r\left(1 - \frac{2M}{r}\right). \end{aligned} \tag{3.19}$$

One possible solution is that both the left-hand terms here are zero; that is, both r and p_r are constant: a circular orbit. We can see from the second line of (3.19), that $\dot{r} = 0$ with r finite gives $p_r = 0$. Writing the constant value of r as a, we then have the circular orbit description:

$$r = a, \quad p_r = 0, \quad \dot{\phi} = \frac{p_\phi}{a^2}, \quad p_\phi^2 = \frac{a^3 M}{(a-2M)^2}. \tag{3.20}$$

Interpreting the individual terms in Eqs. (3.19) under these conditions, we see that p_ϕ^2/r^3 pushes the orbiting particle *outwards*, while $-M/(r-2M)^2$ pulls it *inwards*, and for a circular orbit both balance. The interesting question then becomes, is this orbit stable?

To see why it might be *un*stable, consider what happens to $\dot{p}_r$ (which in Newtonian terms would be a force the particle feels; in relativistic terms, it is just a change of momentum) if the circular orbit is perturbed slightly by Δr:

$$\Delta\dot{p}_r = \left(\left.\frac{d\dot{p}_r}{dr}\right|_{r=a,p_r=0}\right)\Delta r = \frac{(6M-a)M}{a(a-2M)^3}\,\Delta r\,. \tag{3.21}$$

Consider the direction of $\Delta\dot{p}_r$ at various radii outside the event horizon. Note how the expression on the right changes sign at $a = 6M$. If $a > 6M$, $\Delta\dot{p}_r$ has the opposite sign to Δr. As a result, any perturbation Δr will cause a Δp_r, and therefore a $\Delta\dot{r}$, in the opposite direction, thereby maintaining quasi-circular orbits. When starting with $a < 6M$, however, $\Delta\dot{p}_r$ and the resulting $\Delta\dot{r}$ will have the same sign as Δr, thus reinforcing the perturbation and driving the particle away from its circular orbit. Hence, stable circular orbits exist only for $a \geq 6M$. Inward of that, circular orbits are unstable and prone to falling into the horizon. The boundary case circular orbit with $a = 6M$ has its own descriptive name: the innermost stable circular orbit (ISCO).

The presence of the ISCO is especially relevant when there is gas in the vicinity of a black hole. Drag forces cause gas and dust to turn some of their orbital kinetic energy into heat, which is then radiated away. Loss of kinetic energy in this way means that particle orbits in the vicinity of a massive body (not necessarily a black hole) can change from unbound to bound. That is, gas which was set to fly past the massive body instead falls into orbit around it. Energy continues to be radiated away, with the orbiting particles falling ever deeper into the gravitational field. Consider the Earth: in the units of this chapter, we could say that it is in a near-circular orbit with $a \sim 10^8 M$ and a specific orbital energy (from Eq. (1.29)) about $-M/2a$. To have particles go from an unbound energy state to that radius means that the gas and dust that eventually formed the Earth must have radiated away $\sim 10^{-8}$ of an Earth-mass of energy in order to form the planet at the present location. There is—fortunately for us—not much drag on the Earth now, but if there were lots of gas nearby, the Earth would get dragged into ever-decreasing circles around the Sun, radiating away ever more orbital energy, until it got devoured by the Sun. Meanwhile, the orbital energy down to the solar radius (amounting to $\sim 10^{-6}$ of the mass) would get radiated away. The process of a compact object, such as a star or a black hole, collecting more mass in this way is known in astrophysics as *accretion*. Though we will not cover it in this book, accretion is a favourite topic for numerical simulations—for a nice example, see APOD `050312`. In the case of black holes, accretion occurs with gas going on ever smaller circular orbits down to the ISCO, and then falling into the horizon as the orbits become unstable. Orbital energy continues to get radiated away all the way down to the ISCO. The accretion process means that, paradoxically, black holes are an energy source.

Exercise 3.3 Calculate the value of H for a circular orbit of radius a. Verify that the orbital energy is $-1/(2a)$ for large a. At the ISCO, show that the orbital energy is $-1/16$. (This number is an upper limit on mass–energy conversion through accretion, and can be interpreted as the efficiency of a black hole.) Finally, find a for light in a circular orbit.

3.5 Cartesian Variables

So far we have mainly worked in spherical polar variables within the Schwarzschild space. For some purposes, especially numerical integration, Cartesian variables are more convenient. They are also useful for discussing the topic of the next section, gravitational lensing. As noted in the previous section, once in orbit around a central Schwarzschild mass, a particle remains within a given orbital plane, so that we only need to be concerned with describing the system in two spatial dimensions.

The relevant canonical transformation formulas to go from (r, ϕ) to (x, y) and from (p_r, p_ϕ) to (p_x, p_y) are used fairly often in Hamiltonian dynamics, and they are given in Eqs. (B.9) and (B.11). Substituting these into the Hamiltonian (3.17) and simplifying leads to

$$H(x, y, p_x, p_y) = \tfrac{1}{2}\left(p_x^2 + p_y^2\right) - \frac{r/2}{r - 2M} - M\frac{(xp_x + yp_y)^2}{r^3}. \qquad (3.22)$$

Here r is not to be considered a coordinate in its own right, but simply an abbreviation for $\sqrt{x^2 + y^2}$. The relation between coordinate time and proper time in Eq. (3.14) remains the same.

In books on general relativity you will also find variant forms of the Schwarzschild metric and the implied Hamiltonian, which are associated with different choices of 'gauge'. These arise, essentially, because in curved space it is possible to define x, y, z or r in slightly different ways. This chapter follows the so-called *standard gauge.*

Exercise 3.4 Numerically integrate Hamilton's equations for the Schwarzschild Hamiltonian in two dimensions using Eq. 3.22 and find some interesting orbits. A useful first case to try is to choose the initial value $(x = a, y = 0, p_x = 0, p_y = \sqrt{M/a})$ for some $a \gg M$. Some of the derivatives you will need are given in the following section.

3.6 Gravitational Lensing

Orbits under the Schwarzschild Hamiltonian (3.22) describe the trajectories of light if $H = 0$. These can be unbound orbits, analogous to the Newtonian unbound orbits we studied in Section 1.6. In fact, the most common regime for photons involves orbits that are so unbound, they barely differ from a straight line.[7] In other words, we are not so interested in the central masses 'capturing' anything (as we were with protoplanets capturing material in Section 1.6), but instead we are focused on describing the deflection of light rays under gravity. This phenomenon is known as *gravitational lensing* by analogy with optics. Deflections of light rays do not have to be large to be observable, as positions of objects on the sky are routinely measured to accuracies of a micro-radian (< 1 arc-second).

Historically, light deflection was the first prediction of general relativity to be tested, by a famous set of observations by Dyson, Eddington, and Davidson in 1919,[8] which appears to have been the key event that first made Einstein a household name.

Let us now consider the perturbation of a straight-line orbit in the Schwarzschild spacetime, which may correspond to light ($H = 0$), or a slow-moving particle (when H is close to $-\frac{1}{2}$), or anything in between. In these cases the Hamiltonian (3.22) gives us

$$\begin{aligned}\frac{\dot{p}_x}{M} &= -\frac{x}{r(r-2M)^2} + \frac{2p_x(xp_x + yp_y)}{r^3} - \frac{3x\,(xp_x + yp_y)^2}{r^5}, \\ \dot{y} &= p_y - 2M\,\frac{y\,(xp_x + yp_y)}{r^3},\end{aligned} \qquad (3.23)$$

and analogous expressions for $\dot{p}_y$ and $\dot{x}$. Orbits that have $r \gg M$, which are therefore only slightly affected by M, will be nearly straight. In particular, say that the orbit

[7] Note that in following a geodesic, the photon always travels along a locally 'straight line'. Here, we refer to how an outside observer would claim the trajectory looked.

[8] It is sometimes stated that 'later examination of the photographs taken on that expedition showed the errors were as great as the effect they were trying to measure'. In fact, digitization and re-analysis of the surviving photographs 60 years later supported the original result. (Kennefick, 2012) is a fascinating article separating fact and fiction surrounding the report of the initial measurement.

is approximately parallel to the y-axis and (following the language of Section 1.6) has an impact parameter $b \gg M$. Mathematically, this orbit then has

$$x \simeq b, \qquad p_x \simeq 0, \qquad \dot{y} \equiv \frac{dy}{d\tau} \simeq p_y\,, \tag{3.24}$$

with p_y nearly constant but dependent on the speed. For a slow-moving body, $|p_y|$ will also be small. For light, the condition that $H = 0$ will lead to $p_y^2 \simeq 1$, since $b \gg M$ makes all the r-dependent terms in H very small. Assuming that the trajectory is moving from negative to positive y, p_y will be positive, and we have

$$p_y \begin{cases} \ll 1 & \text{for slow bodies,} \\ \simeq 1 & \text{for light.} \end{cases} \tag{3.25}$$

'Deflection' means picking up some component of motion orthogonal to the current trajectory, that is, picking up some Δp_x, while passing by the mass M. We are therefore interested in how much change of momentum in the x-direction occurs during the entirety of the journey. This is quantified by the integral

$$\Delta p_x = \int_{-\infty}^{\infty} \dot{p}_x\, d\tau = \int_{-\infty}^{\infty} \frac{\dot{p}_x}{p_y}\, dy \simeq \frac{1}{p_y} \int_{-\infty}^{\infty} \dot{p}_x\, dy\,, \tag{3.26}$$

where we have used the 'nearly straight' property (3.24) to change the integration variable from τ to y, and we approximate the momentum in the y-direction as remaining roughly constant.

The expression for $\dot{p}_x$, which is needed to solve the deflection integral, above, is given in Eq. (3.23). However, we can simplify it with the help of the small-deflection conditions in Eqs. (3.24)–(3.25), so that

$$\frac{\dot{p}_x}{M} \simeq \begin{cases} -\dfrac{b}{(b^2+y^2)^{3/2}} & \text{for slow bodies,} \\ -\dfrac{b}{(b^2+y^2)^{3/2}} - \dfrac{3by^2}{(b^2+y^2)^{5/2}} & \text{for light.} \end{cases} \tag{3.27}$$

Each of the terms in (3.27), when integrated over y, contributes $-2/b$. Hence we have

$$\Delta p_x \simeq \begin{cases} \dfrac{2M}{bp_y} & \text{for slow bodies,} \\ \dfrac{4M}{b} & \text{for light.} \end{cases} \tag{3.28}$$

The slow-body formula is the Newtonian result (and for light, recall that $p_y \simeq 1$ here). Thus, relativity leads to this Schwarzschild gravitational lens deflecting light twice as much as one would expect by using the slow-body formula or Newtonian regime.

A further, and remarkable, phenomenon occurs when the deflection angle is more than the angular size of the massive body. If the body has radius R and sits at a distance D, then its apparent radius on the sky will be R/D. Light from a more distant object, if it passes nearby, will be deflected by $4M/R$ according to Eq. (3.28). If $4M/R > R/D$ an observer can receive deflected light from opposite sides of the body; that is, one can observe two or even more images of the same source. This phenomenon is known as gravitational lensing. Several examples of galaxies and clusters of galaxies acting as gravitational lenses are known, such as in APOD `160828`, APOD `170506`, and even a smiling example APOD `151127`. Note that the condition $4M/R > R/D$ is more likely to be fulfilled in situations where D is large. This is the reason most known gravitational lenses are distant galaxies or clusters of galaxies. Being distant, gravitationally lensed systems are also faint, which explains why none were known until Walsh, Carswell, and Weymann discovered the first example in 1979.

Gravitational lenses are not precisely configured like camera lenses or our own eye lenses; they are more analogous to irregular beads of glass. In particular, gravitational lenses do not have a focal length. There is, however, a different quantity, known as the *Einstein radius*. To understand it, suppose the lensing mass is perfectly round, and that the background light source is located precisely behind it and much much further away, so that the light rays are almost parallel before deflection. The observer can then see light that has been deflected to us from all around the lensing body. The result is a ring of light, known as an *Einstein ring*. The radius of the ring on the sky will then be simply the deflection angle, which we can solve for as $\sqrt{4M/D}$. This is the Einstein radius. The perfect alignment of a light source exactly behind a lensing mass is very unlikely, and, even in our large Universe, no perfect Einstein rings have yet been observed. However, several approximate yet beautiful Einstein rings are known: good examples are APOD `111221` and APOD `160420`.

Exercise 3.5 The Einstein radius as described in this section is for the case where light rays are parallel before being deflected. In other words $D_S \gg D$, where D_S is the distance to the source. Try and generalize to smaller D_S.

---

4 Interlude: Quantum Ideal Gases

At the end of the 19th century, one of the leading astronomers of the day, Simon Newcomb, made an unintentionally ironic statement:[1] 'We are probably nearing the limit of all we can know about astronomy.' Newcomb was no fool—among his achievements was measuring the mass of Jupiter correctly to five digits using its gravitational perturbation of asteroid orbits. But the remark reminds us that the character of astrophysics changed around 1900. Celestial mechanics, after dominating astrophysics for two centuries, faded in popularity, until it was partly revived with the coming of space travel. Even Poincaré's discovery of deterministic chaos lay almost unnoticed for decades, until computers became common.

In place of gravitational dynamics, microphysical quantum mechanics and its macrophysical consequences took centre stage. In 1900, the macroscopic properties of the Sun (such as mass and temperature) were fairly well-quantified and were starting to be measured for nearby stars. But the interior properties of a star could only be speculated about, and the source of its continued energy production was a complete mystery. The situation started changing when a theory of radiative processes arrived. It became possible to calculate the internal structure of a star, even without knowing what the energy source was, using the known macroscropic properties as constraints. When the underlying energy source was eventually identified as nuclear fusion in the 1930s, a full theory of stellar structure and evolution developed quite quickly.

The next part of this book is about the physics of stars and their structure. The description as 'balls of gas, burning, billions of miles away' from *The Lion King* is not completely accurate, but not a bad place to begin, and in this chapter we will develop the concept and statistics of ideal gases. Classically, an ideal gas is a substance made of many, many identical particles moving freely in some region, with their only relevant microscopic properties being (a) their number density and (b) how fast they are going, or more precisely, their momenta. While the particles are *identical* in the sense of being indistinguishable, they can have very different states and a wide range of momenta, which are describable by statistical distributions. In these ideal classical gases, macroscopic properties like temperature and pressure all derive from the underlying spatial and momentum distributions.

A *quantum* ideal gas follows the same general idea, except that the distributions describing the properties reveal that the particles behave very differently from our everyday intuition, because: (i) the momentum values of the particle states are quan-

[1] In fairness to Newcomb, we could not trace any contemporary source for this comment, even though it is widely quoted.

The Astronomer's Magic Envelope. Prasenjit Saha and Paul A. Taylor.

10.1093/oso/9780198816461.001.0001

tized, and (ii) the particle speeds may be relativistic. Stars are predominantly ionized hydrogen and helium, but the electrons and nuclei behave like coexisting classical ideal gases. Thermal radiation is an example of an ideal gas where quantum behaviour is essential.

4.1 Planckian Units

When quantum and thermodynamic processes first appeared in astrophysics, two new physical constants also appeared, joining the constants G and c. These were, respectively, the reduced Planck constant $\hbar$ and the Boltzmann constant k_B. We can make the task of keeping track of these constants much easier by a good choice of units. In previous chapters we made use of the freedom to choose units. While discussing the gravitational constant in Chapter 1, we saw how using different units of length and time can make a problem cleaner. For the three bodies in Chapter 2 we basically defined our own units of mass, length, and time, specially adapted to the problem. For the Schwarzschild spacetime, we chose customized units for length and time.

A key idea is to let the universal constants themselves define the units. In the research literature, one often finds a statement like

$$c = \hbar = G = k_B = 1 . \tag{4.1}$$

This is just scientific shorthand for saying that new units of mass, length, time, and temperature are being introduced, such that the numerical values of c, $\hbar$, G, and k_B are all unity. With dimensions written out, equation (4.1) says:

$$\begin{aligned} c &= 1\,\mathrm{lap\,tick}^{-1} , \\ \hbar &= 1\,\mathrm{marb\,lap}^2\,\mathrm{tick}^{-1} , \\ G &= 1\,\mathrm{marb}^{-1}\,\mathrm{lap}^3\,\mathrm{tick}^{-2} , \\ k_B &= 1\,\mathrm{marb\,lap}^2\,\mathrm{tick}^{-2}\,\mathrm{therm}^{-1} . \end{aligned} \tag{4.2}$$

Actually, the new units of mass, length, time, and temperature have no standard names, so for the purposes of this text we have simply bestowed upon them the representative and illustrative names of marble (marb), lap, tick, and therm. Rearranging the definitions (4.2) gives the new units in terms of the universal constants, and hence their values in SI:

$$\begin{aligned} \mathrm{marb} &\equiv (\hbar c/G)^{1/2} = 2.177 \times 10^{-8}\,\mathrm{kg} , \\ \mathrm{lap} &\equiv \left(\hbar G/c^3\right)^{1/2} = 1.615 \times 10^{-35}\,\mathrm{m} , \\ \mathrm{tick} &\equiv \left(\hbar G/c^5\right)^{1/2} = 5.383 \times 10^{-44}\,\mathrm{s} , \\ \mathrm{therm} &\equiv \left(\hbar c^5/G\right)^{1/2} k_B^{-1} = 1.419 \times 10^{32}\,\mathrm{K} . \end{aligned} \tag{4.3}$$

Units defined in this way are known as Planckian units.[2] The definition is not standardized, and varies slightly across the literature. In particular, some authors use h

[2]Planck himself just called them natural units, and argued that even extraterrestrial cultures would view them as 'natural' choices—while, as noted earlier, SI units arose from consideration of specifically human scales.

rather than $\hbar$, which makes all the units larger by $\sqrt{2\pi}$. Some further quantities and conversions are described in Appendix D.

We note that a mass of 1 marb is small but at least measurable with existing technology. However, laps and ticks are both incredibly tiny even compared to the smallest scales in particle physics. A therm, on the other hand, is hotter than hot; the energy per particle at a temperature of 1 therm would be of order a marb, and that is many orders of magnitude beyond the most energetic known particles (ultra-high-energy cosmic rays). Nonetheless, Planckian units do allow for some tremendous simplifications in equations, because as the shorthand Eq. (4.1) suggests, cumbersome and distracting factors of $\hbar, c, G, k_{\rm B}$ in strange powers just vanish from terms. Those factors are not really lost, however, and can be directly summoned back whenever desired, as we will see below. Nor does using Planckian units imply abandoning SI units. In fact, it is *vital* to be able to switch comfortably between Planckian and SI units, and we will do so whenever we need to compare theory and measurement, or simply wish for some intuition on human scales.

It is interesting that SI units themselves are gradually evolving to being defined in terms of universal constants, while keeping the values people-sized as before. In the 1960s, the metre was defined in terms of the wavelength of a spectral line of krypton, and the second in terms of a characteristic frequency of caesium atoms. The definition of the second still stands, but since 1983, the metre has been defined as $1/299,792,458$ of a light-sec in a vacuum. As of 2017, the standard kilogram is still a metal cylinder made in 1879, but a new definition in terms of Planck's constant

$$\mathrm{kg} = \frac{2\pi\hbar}{6.62606(\ldots)\times 10^{-34}}\,\mathrm{m}^{-2}\,\mathrm{s} \tag{4.4}$$

is in process. The exact number is yet to be chosen, and the desired criterion is that old and new kilogram match to eight digits. Similarly, defining a degree Kelvin as $1.38065(\ldots)\times 10^{-23} k_{\rm B}^{-1}\,\mathrm{kg\,m^2\,s^{-2}}$ is being considered. Thus, three-quarters of Planck's definition are on their way to be adopted. One can imagine going even further and defining the second so that the gravitational constant is given exactly by an expression such as Eq. (1.12). But, as noted in Chapter 1, calibrating against G to eight significant digits remains just a dream at present, due to the relative weakness of gravity compared to other forces.

Three particular microphysical quantities will play a key role in the rest of this book. These are the electron mass m_e, the baryon mass $m_{\rm b}$ (which we take as the mean of the proton and neutron masses), and the fine structure constant α (which is dimensionless). These have values

$$\begin{aligned} m_e &= 4.184\times 10^{-23}\,\mathrm{marb}\,,\\ m_{\rm b} &= 7.688\times 10^{-20}\,\mathrm{marb}\,,\\ \alpha &\equiv \frac{1}{4\pi\epsilon_0}\frac{e^2}{\hbar c} = \frac{1}{137.036}\,, \end{aligned} \tag{4.5}$$

where e is the charge of one electron.

Let us now see how Planckian units work in some examples. As a first example, consider the Bohr radius, which is the scale of atomic sizes. In Planckian units a Bohr radius is given by

$$r_{\rm B} = (\alpha m_e)^{-1}\,. \tag{4.6}$$

As written, this quantity currently has a numerical value of 3.275×10^{24} and units of marb^{-1}, therefore looking like an inverse mass in Planckian units. If we wish to interpret it as a length, we simply multiply by marb $\times$ lap to 'cancel out' the mass property and 'put in' length. Since by (4.3) a lap marb $= \hbar/c$ in terms of universal constants, through multiplication we recover the SI formula $r_{\rm B} = \hbar/(\alpha\, c\, m_e)$. If we want the numerical value of $r_{\rm B}$ in metres, that is even simpler: we just take the numerical value of $r_{\rm B}$ in Planckian units and multiply by a lap, giving us 5.29×10^{-11} m. Hydrogen has an atomic radius of $r_{\rm B}$, helium is smaller, lithium $\simeq 3r_{\rm B}$, oxygen $\simeq r_{\rm B}$ again, and so on.

As a second example, consider the electrostatic potential between two nuclei with atomic numbers Z_1 and Z_2 separated by distance r. In Planckian units, the potential is written as

$$V(r) = \frac{\alpha Z_1 Z_2}{r}\,. \tag{4.7}$$

To turn this apparent inverse length into an energy, we multiply by lap because of the $1/r$, and then further by marb lap^2 tick^{-2} to get the dimensions of energy. The result is $\hbar c$, giving an expression for the Coulomb potential in SI units, $V = \alpha\hbar c\, Z_1 Z_2/r$. To get the numerical value in SI units, we take the numerical value in Planckian units and multiply by marb lap^2 tick^{-2}.

Exercise 4.1 The binding energy of an electron in a hydrogen atom (equivalently, the ionization energy) equals $\frac{1}{2}\alpha^2 m_e$ and is called a Rydberg and can also be written as $\alpha/2r_{\rm B}$. Calculate the wavelength in microns of a photon with this energy. That wavelength is known as the Lyman edge.

Exercise 4.2 Sometimes we are confronted with some specialised non-SI unit and need to state results in terms of it. In these situations, the strategy is to find a hook (or crook) that lets us convert to and from the home (Planckian) system of units. A good example is the electron-volt (eV), which is commonly used in particle physics as a unit of mass or energy. A convenient working definition (though not the formal definition) of an eV is that the electron mass $m_e = 512\,$keV, where keV stands for kilo electron-volt or $10^3\,$eV.

Evaluate an eV in Planckian units. Then verify that a Rydberg is $13.6\,$eV and that room temperature $\simeq 0.025\,$eV.

Exercise 4.3 Verify that when $\frac{1}{2}(\alpha/m_e)^3$ is interpreted as a time, it is about the age of the Earth, and work out the factor needed to construct a time in non-Planckian units. This particular example is merely an amusing coincidence, but in later chapters we will meet similar-looking expressions which really do have physical meanings as time scales. Perhaps you can invent other amusing or thought-provoking examples.

4.2 Phase-Space Distributions

We noted in the introductory part of this chapter that the key microscopic property of a gas is the distribution of particles in space and in momentum. This can be formalized and expressed quite generally by a distribution function

$$f(\mathbf{r}, \mathbf{p}), \tag{4.8}$$

whose domain is the six-dimensional phase space of position $\mathbf{r}$ and momentum $\mathbf{p}$. For a given type or class of particle, this function theoretically describes the relative number of particles with a given position and momentum.

Energy and velocity are functions of the momentum, assuming the mass is known. This is true even if the particles are relativistic. In Planckian units

$$E = \sqrt{m^2 + p^2}, \qquad \mathbf{v} = \frac{\mathbf{p}}{E}. \tag{4.9}$$

We remark that the two equations here can be rearranged to express E and $\mathbf{p}$ in terms of $\mathbf{v}$, as

$$E = \frac{m}{\sqrt{1 - v^2}}, \qquad \mathbf{p} = \frac{m\mathbf{v}}{\sqrt{1 - v^2}}, \tag{4.10}$$

which may be familiar from special relativity. In phase space, however, the expressions (4.9) in terms of $\mathbf{p}$ are more useful. There are two important limiting regimes. The non-relativistic regime is where $p \ll m$, which gives

$$E = m + \frac{p^2}{2m}, \qquad \mathbf{v} = \frac{\mathbf{p}}{m}. \tag{4.11}$$

The extreme relativistic regime is where $p \gg m$, in which

$$E \approx p, \qquad v \approx 1. \tag{4.12}$$

Integrating f over the momentum components yields the number density

$$n(\mathbf{r}) = \int f(\mathbf{r}, \mathbf{p})\, d^3\mathbf{p} \tag{4.13}$$

in ordinary space. A further integration over space $n = \int n(\mathbf{r})\, d^3\mathbf{r}$ would give the total number of particles in the volume. Let us, however, not integrate over space but instead focus attention on some given $\mathbf{r}$ and its neighbourhood. For the rest of this chapter, we will omit writing the $\mathbf{r}$ dependence explicitly, but it is understood that all gas properties depend on location.

Still considering properties of momentum, we now make the simplifying assumption that f is isotropic; that is, there is no preferred direction. Strictly speaking, this is untrue in a star, because the local gravitational field defines a special direction. Let us assume, however, that we are in a neighbourhood of $\mathbf{r}$ so small that the difference in gravitational potential from top to bottom is negligible compared to particle energies. Under isotropy, f depends on the magnitude p of momentum, but not its direction.

It is useful then to introduce polar coordinates for the momentum, which relate to Cartesian components as

$$(p_x,\ p_y,\ p_z) = (p\sin\theta\cos\phi,\ p\sin\theta\sin\phi,\ p\cos\theta)\,. \tag{4.14}$$

As with spatial polar coordinates, magnitude $p \geq 0$, polar component $\theta \in [0,\ \pi]$, and azimuthal component $\phi \in [0,\ 2\pi]$. (We emphasize, however, that θ and ϕ here are not spatial coordinates; they are simply angles giving the direction of the momentum. They should also not be confused with the momentum variables p_θ and p_ϕ appearing in Hamiltonian dynamics.) A differential element in momentum space can now be written as

$$f(\mathbf{p})\,d^3\mathbf{p} = f(p)\,p^2\,dp\,d\Omega, \qquad d\Omega \equiv \sin\theta\,d\theta\,d\phi\,. \tag{4.15}$$

Here $d\Omega$ is a differential solid angle, and integrating over the allowed ranges of θ and ϕ, one finds that $\int d\Omega = 4\pi$. Hence, in a system with an equal orientational probability of a given momentum, the spatial number density is

$$n = 4\pi \int f(p)\,p^2\,dp\,. \tag{4.16}$$

Distributions are useful in describing other macroscopic quantities of interest as well. For example, the kinetic energy of a particle is always a function of its momentum, so that $E = E(p)$. The kinetic energy density is therefore obtained by integrating over E with the distribution as follows:

$$u = 4\pi \int E(p)\,f(p)\,p^2\,dp\,. \tag{4.17}$$

Although we often treat $f(\mathbf{p})$ as having no dependence on θ or ϕ, we may still be interested in macroscopic properties referring to a particular direction. For example, imagine a gas in a container with a hole in it, and a vacuum outside. Particles will leave at the rate $nv_\perp$ times the area of the hole, where $v_\perp$ is the velocity perpendicular to the hole. If we take $\hat{\mathbf{z}}$ as the direction perpendicular to the hole, then $v_\perp = v_z = v(p)\cos\theta$, where $v(p)$ is the speed of a particle with momentum p. The outflow rate per unit area of the hole will be

$$\Phi = \int v(p)\ \cos\theta\,f(p)\,p^2\,dp\,d\Omega\,. \tag{4.18}$$

The integral over Ω does not involve $v(p)$ or $f(p)$, and hence it can be separated and carried out immediately. The result is

$$\Phi = \pi \int v(p)\,f(p)\,p^2\,dp\,. \tag{4.19}$$

If we are interested in the rate of energy outflow rather than particle outflow, we simply introduce a factor of energy per particle $E(p)$ inside the integral. The result is the radiant intensity

$$I = \pi \int v(p)\,E(p)\,f(p)\,p^2\,dp\,, \tag{4.20}$$

or energy outflow rate per unit area.

Another modification of the particle outflow rate leads to an expression for gas pressure. Imagine covering up the hole, so that gas particles bounce off the cover instead of leaving. Each bounce reverses the momentum component $p_\perp$ perpendicular to the cover, giving it a change momentum equal to $2p_\perp = 2p\cos\theta$. The cover therefore receives momentum per unit area at the rate

$$P = \int (2p\cos\theta)\, v(p)\, \cos\theta\, f(p)\, p^2\, dp\, d\Omega \tag{4.21}$$

which is otherwise known as a pressure. Carrying out the Ω integral gives

$$P = \frac{4\pi}{3} \int p^3\, v(p)\, f(p)\, dp\,. \tag{4.22}$$

4.3 Quantum Many-Particle Distributions

In order to evaluate the integrals over p in the previous section, it is necessary to know the form of $f(p)$. We have already stated that we are working within the realm of the simplifying assumptions of the ideal-gas approximation—namely, that the particles are point-like and do not interact with each other. The particles may, however, be relativistic (so that v is of order 1), and quantum properties may be important. In the domain of quantum mechanics, the phase-space density then takes one of two well-known forms based on the denominator in the expression:

$$f_{\rm QM}(p) = \frac{g}{(2\pi)^3} \frac{1}{e^{(E(p)-\mu)/T} \pm 1}\,, \tag{4.23}$$

each of which approaches the same classical limit (described in the next section). There are several ways to derive this expression, and they are found in most introductions to statistical mechanics. So we will not reproduce a derivation here, and instead just explain the symbols.

For the plus sign, $f_{\rm QM}(p)$ applies to Fermi–Dirac particles (fermions), such as electrons and baryons; for the minus sign, $f_{\rm QM}(p)$ applies to Bose–Einstein particles (bosons), such as photons. The $(2\pi)^3$ term in the denominator comes about from the quantization of phase space into finite cells. We should really be summing over cells in phase space, but instead we are integrating and then dividing by the volume of each cell, which is h^3 in ordinary units and $(2\pi)^3$ in Planckian units. The integer constant g, known as the *degeneracy*, is the number of allowed quantum states per cell, and equals the number of possible internal states of the particle. For electrons, protons, neutrons, and photons $g = 2$, corresponding to two spin states. But for atoms and molecules, g is larger, because of possible rotation and vibration modes for more complicated structures. The parameter T is the temperature. Finally, there is the parameter μ, which behaves like a reference point for E. It is the energy associated with introducing a new particle into the system. For historical reasons, μ is called the *chemical potential.*

4.4 The Classical Ideal Gas

If $e^{(E-\mu)/T} \gg 1$, the phase-space density (4.23) reduces to

$$f(p) = \frac{g}{(2\pi)^3} e^{-(E(p)-\mu)/T} . \tag{4.24}$$

This can be interpreted as the case in which phase space is so sparsely populated that quantization is not noticeable. Nor is the distinction between bosons and fermions noteworthy, as the particles have so much more energy compared to the lowest states where their behaviour differs. Assuming the particles are non-relativistic (see Eq. (4.11)), we then have what is known as the Maxwell–Boltzmann distribution:

$$f_{\rm MB}(p) = \frac{g\, e^{(\mu-m)/T}}{(2\pi)^3} e^{-p^2/2mT} , \tag{4.25}$$

which describes the phase-space distribution of a classical ideal gas. From the phase-space distribution, we will now derive thermodynamic quantities.

To obtain the pressure and energy density, let us first write down the integral identity

$$\int_0^\infty p^4 e^{-p^2/2mT}\, dp = 3mT \int_0^\infty p^2 e^{-p^2/2mT}\, dp , \tag{4.26}$$

which can be derived by integrating the left-hand side by parts. We then rewrite this identity, first by comparing with the phase-space density in Eq. (4.25), and then by expressing the left-hand side in two slightly different ways, to obtain

$$2\int_0^\infty \frac{p^2}{2m} p^2 f_{\rm MB}(p)\, dp = \int_0^\infty p^3 \frac{p}{m} f_{\rm MB}(p)\, dp = 3T \int_0^\infty p^2 f_{\rm MB}(p)\, dp . \tag{4.27}$$

Now recall (e.g., from Eq (4.11)) that for non-relativistic particles, $p^2/2m$ is $E(p)$ not including the rest mass, while p/m is $v(p)$. Hence the left-hand side of Eq. (4.27) is proportional to the internal energy in Eq. (4.17), and the middle is proportional to the pressure in Eq. (4.22). Meanwhile the right-hand side is proportional to the number density in Eq. (4.16). Putting in the various coefficients, we get

$$2u = 3P = 3Tn , \tag{4.28}$$

recovering the usual expressions for a classical mono-atomic ideal gas:

$$u = \tfrac{3}{2} nT , \qquad P = nT . \tag{4.29}$$

Finally, let us relate the chemical potential μ to the macroscopic quantities. To do this, note that while there are several terms in Eq. (4.25), only the exponential factor depends on the momentum p. Noting that $p^2 = p_x^2 + p_y^2 + p_z^2$ and using the standard result for a Gaussian integral that

$$\int_{-\infty}^\infty e^{-x^2}\, dx = \sqrt{\pi} , \tag{4.30}$$

we can integrate $f_{\rm MB}$ over all $\mathbf{p} = (p_x, p_y, p_z)$ as in Eq. (4.13) to obtain the number density

$$n_{\rm MB} = \left(\frac{mT}{2\pi}\right)^{3/2} g\, e^{(\mu-m)/T} . \tag{4.31}$$

We will need this relation in the last chapter of this book, when we consider atomic versus ionized hydrogen at earlier epochs of the Universe.

Exercise 4.4 Imagine a Maxwell–Boltzmann gas with number density n in a container. A hole is punched in the container, through which gas molecules leak into a vacuum outside. Show that the leakage rate is $n\sqrt{T/2\pi m}$ times the area of the hole.

4.5 A Photon Gas

Photons are bosons with two spin states (or degeneracy $g = 2$), and no rest mass (hence $E = p$ in Planckian units and $\mu = 0$). Inserting this information into the phase-space density (4.23), and adopting the subscript γ for photons, we have

$$f_{\mathrm{BE}}(p) \rightarrow f_\gamma(p) = \frac{1}{4\pi^3} \frac{1}{e^{p/T} - 1} . \tag{4.32}$$

For any integrals over p involving this distribution we invoke the identity[3]

$$\int_0^\infty \frac{x^s\, dx}{e^x - 1} = s! \sum_{k=1}^{\infty} \frac{1}{k^{s+1}} . \tag{4.33}$$

For some values of s, the sum has an exact expression, such as $\sum_k k^{-4} = \pi^2/90$, but in other cases it must be evaluated numerically. The integral in Eq. (4.16) for the number density would be a case where $s = 2$ in Eq. (4.33) and comes to

$$n \simeq 0.2435\, T^3 . \tag{4.34}$$

The integrals in Eqs. (4.17), (4.20), and (4.22) for the internal energy, radiant intensity, and pressure, respectively, are all proportional to $\int p^3 f(p)\, dp$. Therefore, all exhibit the same dependence on temperature, merely with different constant terms:

$$u_\gamma = \frac{\pi^2}{15} T^4 , \qquad I_\gamma = \frac{\pi^2}{60} T^4 , \qquad P_\gamma = \frac{\pi^2}{45} T^4 . \tag{4.35}$$

Thermal radiation, regardless of the source, can be described as a photon gas, and it has a distinctive observable signature: its spectrum. We define the spectral energy density $S_p \equiv S(p)$ such that $\int S_p\, dp\, d\Omega = u$. For that we need

$$S_p \equiv E(p)\, p^2\, f(p) . \tag{4.36}$$

Spectra are, however, usually expressed in terms of frequency or wavelength:

$$\nu = \frac{p}{2\pi} , \qquad \lambda = \frac{2\pi}{p} . \tag{4.37}$$

Taking $S_p\, dp = S_\nu\, d\nu = S_\lambda\, d\lambda$ gives

$$S_\nu = 2\pi\nu \frac{2\nu^2}{\exp(2\pi\nu/T) - 1} \tag{4.38}$$

or

$$S_\lambda = \frac{2\pi}{\lambda} \frac{2\lambda^{-4}}{\exp(2\pi/\lambda T) - 1} , \tag{4.39}$$

known as the Planck spectrum. Figure 4.1 shows an example in both S_ν and S_λ forms.

[3]Derivations of this identity can be found in discussions of the Riemann ζ function.

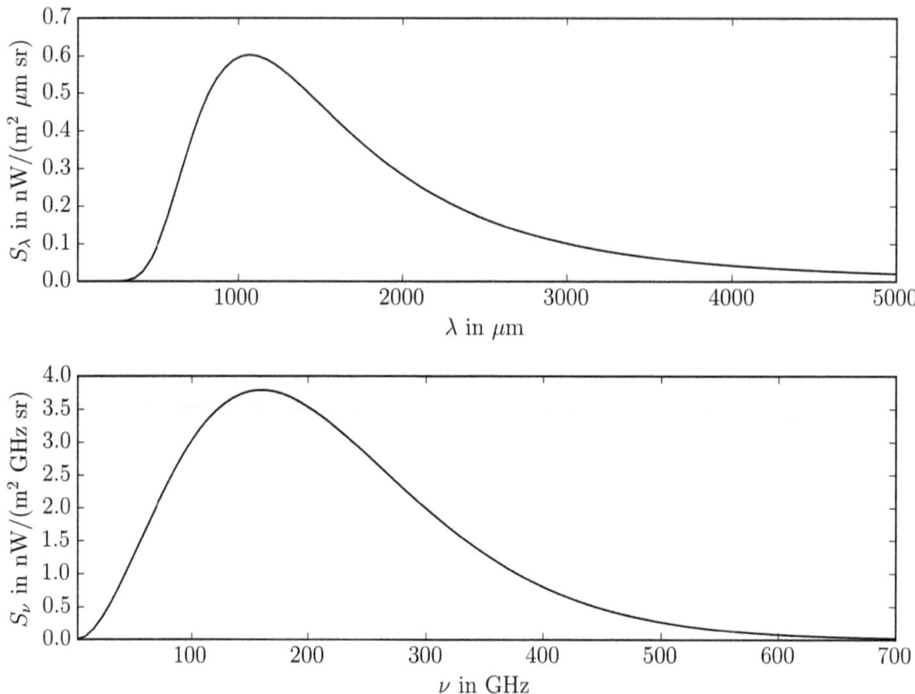

Fig. 4.1 Planckian spectrum at $T = 2.725\,\mathrm{K}$. The units are power (nanowatts, nW) per unit area per steradian per wavelength unit (for S_λ) or per frequency unit (for S_ν). Many possible variations are possible, depending on the exact quantities being plotted. For reference, note that visible light has a range of $\nu \approx$ 400–800 THz ($\lambda \approx$ 375–750 nm).

We can see from the functional form of the Planck spectrum that most of the energy is in the region $p \sim T$, falling off at very small and very large p. The peak frequency and wavelength can be computed by taking $\partial S_\nu/\partial\nu = 0$ and $\partial S_\nu/\partial\lambda = 0$ respectively, which results in

$$\nu_{\mathrm{W}} = \frac{2.821}{2\pi} T\,, \qquad \lambda_{\mathrm{W}} = \frac{2\pi}{4.965} T^{-1}\,. \tag{4.40}$$

The subscript 'W' refers to Wien's displacement law, which states that $\lambda_{\mathrm{W}} T$ is a constant, and was a precursor of Planck's formula. Note that $\nu_{\mathrm{W}} \neq 1/\lambda_{\mathrm{W}}$. This inequality may seem strange, but it is only a consequence of the way S_ν and S_λ have been defined, and has no physical significance.

The Planck spectrum is also known as *blackbody radiation.* A blackbody is a surface that does not reflect; it absorbs all photons landing on it, while itself radiating a Planckian spectrum. Thus, the radiation coming from a blackbody contains no reflected light, but is purely Planckian. Calling this radiation 'blackbody radiation' seems reasonable enough at room temperatures, since ν_{W} and λ_{W} are far in the infrared, and thermal radiation at room temperatures looks pretty dark to us. On the other hand, calling sunlight (which is not far from Planckian) 'blackbody radiation' does seem strange. Perhaps rainbow radiation would be a better term.

Exercise 4.5 The energy density and irradiance of a photon gas are more conventionally written as aT^4 and σT^4, respectively, defining the radiation constant $a = \pi^2/15$ and the Stefan-Boltzmann constant $\sigma = \pi^2/60$ (both in Planckian units). Derive these constants in non-Planckian units, by using physical arguments to multiply by the appropriate powers of marb, lap, tick and therm.

Exercise 4.6 Reconstruct expressions for S_ν and S_λ for a photon gas with h, c, k_{B}.

Exercise 4.7 The solar constant (defined as the solar energy flux above the Earth's atmosphere at normal incidence) is measured as $\simeq 1350\,\mathrm{W\,m^{-2}}$. Estimate its value theoretically, by approximating the solar surface as a blackbody radiator with $\lambda_{\mathrm{W}} = 0.5\,\mu\mathrm{m}$ (visible light) and recalling that apparent size of the Sun on the sky implies that the radius of the Earth's orbit is $\simeq 215$ times the radius of the Sun.

Exercise 4.8 The Universe is a large container permeated by radiation, called the cosmic microwave background (CMB) (the topic of Chapter 9), which behaves as a photon gas with $T = 2.725\,\mathrm{K}$. Calculate λ_{W} in metres and the number density as photons $\mathrm{cm^{-3}}$.

Exercise 4.9 Astrophysical data often has large error bars, but not always. The frequency spectrum of the cosmic microwave background was measured so precisely as Planckian with $T = 2.725\,\mathrm{K}$ by the FIRAS instrument on the COBE spacecraft, that in 1990 J. Mather and collaborators, who designed and built FIRAS, had to exaggerate their error bars by ~ 100 to make them visible. You can find many versions of the FIRAS plot online. The output units are usually $\mathrm{cm^{-1}}$ for frequency, and Megajansky per steradian for the spectrum. (A jansky is defined as $10^{-26}\,\mathrm{W\,m^{-2}\,Hz^{-1}}$.) Plot the spectrum S_ν from Eq. (4.38) using the same T and units as FIRAS.

4.6 A Degenerate Fermi Gas

An even more exotic example of a quantum ideal gas is a degenerate Fermi gas, which typically consists of electrons.

Let us recall the general principle, that when a substance is cooled, particles tend to go into ever-lower energy states. A system of fermions, however, can have only one particle per state (the Pauli exclusion principle). Zero temperature is when all energy states up to a certain level, known as the *Fermi level*, are occupied, and there are no particles above the Fermi level; since particle energy increases monotonically with momentum, we can equivalently think of all momentum states being filled up to a corresponding level called the Fermi momentum level, which we write as p_{F}. The phase-space density is simply

$$f_{\mathrm{FD}}(p) = \begin{cases} \dfrac{g}{(2\pi)^3} & p \le p_{\mathrm{F}}\,, \\ 0 & p > p_{\mathrm{F}}\,. \end{cases} \tag{4.41}$$

The degeneracy g is the number of particle states possible at each momentum level. Such a system is known as a degenerate Fermi gas.

Zero temperature is not necessary for a degenerate Fermi gas, any $T \ll E(p_{\text{F}})$ is low enough. To see why, consider the general distribution function (4.23) for a quantum ideal gas, and put $T \ll |\mu|$. The term $e^{(E-\mu)/T}$ changes sharply from tiny to huge around $E = \mu$, cutting off the distribution at $E = \mu$, thus producing the degenerate distribution function (4.41). This argument tells that the Fermi energy $E(p_{\text{F}})$ is also the chemical potential of a degenerate Fermi gas. The temperature, though much smaller than the Fermi energy, may be quite high from other perspectives.

Putting $g = 2$, which applies to electrons and also (for example) neutrons, the degenerate distribution function (4.41) integrates trivially to give the number density

$$n = \frac{{p_{\text{F}}}^3}{3\pi^2}\,. \tag{4.42}$$

It is evident that the Fermi level is determined physically by the number density. Increasing the density raises the Fermi level. Hence, energy must be supplied to a degenerate Fermi gas if it is to be compressed. This gives the gas a pressure. The pressure originates in the particle momentum, but not in the same way as in a classical gas. Nevertheless, the integral expression in Eq. (4.22) for the pressure remains valid, taking the form

$$P = \frac{1}{3\pi^2}\int_0^{p_{\text{F}}} v(p)\,p^3\,dp\,. \tag{4.43}$$

The integral here requires consideration of relativistic particle speeds, as given in Eq. (4.9). The classical and extreme relativistic limits are simple:

$$P \approx \begin{cases} \dfrac{1.91}{m}\,n^{5/3} & p_{\text{F}} \ll m \text{ (non-relativistic)}\,, \\ 0.77\,n^{4/3} & p_{\text{F}} \gg m \text{ (ultra-relativistic)}\,. \end{cases} \tag{4.44}$$

Exercise 4.10 Show that the general case of degeneracy pressure from Eq. (4.43), connecting the limiting cases in Eq. (4.44), is

$$P = (1+u^2)^{1/2}\left(\tfrac{1}{4}u^3 - \tfrac{3}{8}u\right) + \tfrac{3}{8}\sinh^{-1}u\,, \tag{4.45}$$

where $u = p_F/m$.

5 Gravity versus Pressure

Once quantum mechanics was established, a physical understanding of stars developed fairly quickly. The period 1930–50 was particularly active, and the contributions of S. Chandrasekhar are especially important. Over the next three chapters we will look into the basic workings of stars. Along the way, we will mention other objects such as planets and pulsars as well, though we do not cover these other phenomena in depth (fascinating as they are).

There are not many exactly soluble examples to present in this part of astrophysics, so we will adopt some fairly drastic approximations. For example, this often involves simplifying geometry and structure and working with orders of magnitude. Nonetheless, even very simplified models provide useful insight, especially when written in Planckian units, which are used here throughout.

The main topic of this chapter is hydrostatic equilibrium: many objects can be viewed as fluids ('hydro') that are in a balance ('static equilibrium') of inward and outward forces. In general, the inward pull is supplied by gravity, as all the parts of the astrophysical body attract each other, and the magnitude of the squeezing to be overcome (that is, to prevent inward collapse) mainly depends on the amount of mass present. The physical mechanisms that provide an outward, balancing pressure depend on the type of system and matter present; different microphysics can provide the stabilizing forces for certain ranges of required strengths. Therefore, what occurs on the tiniest scales, such as nanometres, determines the overall structure of the object, and the variety of microphysical phenomena is what provides the zoo of astrophysical objects that we observe.

A striking example is that the microphysically defined mass

$$M_{\mathrm{L}} \equiv m_{\mathrm{b}}^{-2} \tag{5.1}$$

sets the mass scale of stellar objects, in terms of the baryon mass. In particular, the solar mass is of this order $M_\odot \simeq 0.54 M_{\mathrm{L}}$. S. Chandrasekhar, A. S. Eddington, and L. D. Landau all arrived at this mass scale from different directions. Since there already is a Chandrasekhar mass and an Eddington luminosity in astrophysics, we call M_{L} the *Landau mass.*

5.1 Spherical Hydrostatic Equilibrium

In a spherical body, the physical statement that pressure balances gravity can be expressed by a differential equation, as follows. Let $P(r)$ be the outward pressure at radius r, $\rho(r)$ be the local density, and let $M(r)$ be the mass enclosed within r. Now

The Astronomer's Magic Envelope. Prasenjit Saha and Paul A. Taylor.

10.1093/oso/9780198816461.001.0001

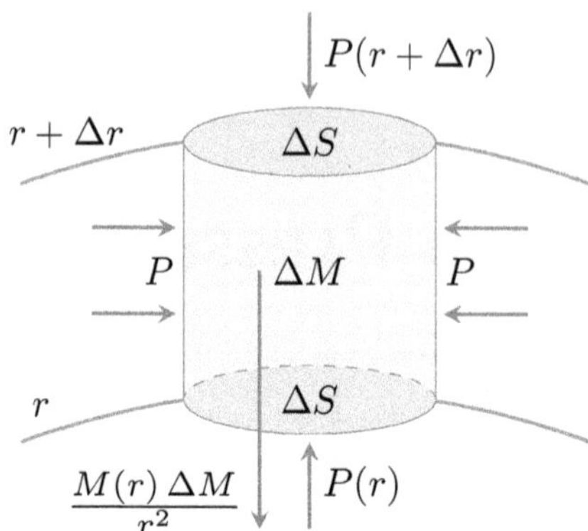

Fig. 5.1 Forces on a mass element of a sphere in hydrostatic equilibrium.

consider a small cylinder as in Fig. 5.1. The cylinder is aligned radially (or 'vertically') and its height is Δr, while its floor and ceiling each have area ΔS. The cylinder will have a downward gravitational force of $\Delta M \times M(r)/r^2$, where ΔM is the mass of the cylinder. (Recall from Section 1.1 that spherical mass at larger r has no net effect.) For equilibrium, the gravitational force must be balanced by the floor to ceiling pressure difference times ΔS. That is to say

$$(P(r) - P(r + \Delta r))\,\Delta S = \frac{M(r)}{r^2}\Delta M. \tag{5.2}$$

For the mass, we have $\Delta M = \rho\,\Delta r\,\Delta S$. Passing to infinitesimals, we get

$$\frac{dP(r)}{dr} = -\frac{M(r)\,\rho(r)}{r^2}, \tag{5.3}$$

known as the equation of hydrostatic equilibrium. For the enclosed mass, we have the usual integral over radius, which we can write as a differential equation

$$\frac{dM(r)}{dr} = 4\pi r^2 \rho(r), \tag{5.4}$$

known as the mass-continuity equation.

Microphysics enters through the relation between P and ρ, known as the *equation of state*. For most substances, the equation of state involves the temperature T as well, a good example being Eq. (4.29) for a classical ideal gas. For such substances, additional differential equations are needed to deal with $T(r)$ in the stellar structure. We will meet these equations in Chapter 7, after studying energy generation in stars. In this chapter, we will confine ourselves to systems where P depends only on ρ.

We remark that spherical hydrostatic equilibrium does not apply to all astrophysical objects: in galaxies, gravity is balanced by centrifugal forces rather than pressure; in asteroids, spherical symmetry does not apply because anisotropic interatomic forces set the shape. Therefore, galaxies and asteroids require other treatments of their equilibrium, which we will not explore in this book.

5.2 Solid Objects: Rock and Ice

The simplest equation of state has a constant ρ, independent of P. Constant density is a useful first approximation for the structure of the Earth and other rocky planets, where interatomic forces maintain a constant density over a wide range of applied pressures. The enclosed mass in Eq. (5.4) is then trivial and the equilibrium condition in Eq. (5.3) is quickly solvable. Note that even though these expressions were derived from considerations of *hydro*static equilibrium, they can still be applied usefully to these 'solid object' cases in which the microphysical forces are approximately isotropic.

The size of an atom is on the order of a Bohr radius $r_{\rm B}$ (see Section 4.1). Molecules are not as tightly packed as atomic sizes, due to the anisotropy of bonds. Taking 4 Bohr radii as the typical intermolecular distance gives a useful microphysical formula for the density. Writing A for the mean mass number, we estimate the density as one atom of mass $A\,m_{\rm b}$ sitting in a cube of edge length $4r_{\rm B}$, yielding

$$\rho \simeq \frac{Am_{\rm b}}{(4r_{\rm B})^3} = Am_{\rm b}\left(\frac{\alpha m_e}{4}\right)^3 . \tag{5.5}$$

Ice or $^1H_2\,^{16}O$ has 18 atomic mass units in three atoms, giving $A = 6$. The designation 'rock' is more heterogeneous and includes many substances. Sand is $^{28}Si\,^{16}O_2$ giving $A = 20$. A somewhat higher value ($A = 30$) reproduces the average density of the Earth, which is understandable since rocks include some heavier elements.

The near-constant density of a solid is maintained by the bulk modulus, which comes from interatomic forces. Recall that bulk modulus K has dimensions of pressure (as do other elastic constants such as Young's modulus), so we can interpret it as an energy per unit volume. Thus K is the number density of atoms, times a characteristic energy per atom, essentially the bond energy. Let us write that

$$\text{bond energy} = \epsilon\, m_e\alpha^2 \tag{5.6}$$

per atom, with $\epsilon \lesssim 1$ depending on the type of bond. The strongest bonds have $\epsilon \simeq 1$, in which case the bond energy is comparable to the ionization energy (recall the Rydberg energy from Section 4.1). Taking the interatomic separation as $4r_{\rm B}$ we can write

$$K \approx \frac{\epsilon\, m_e\alpha^2}{(4r_{\rm B})^3} = \frac{\epsilon}{4^3} m_e^4\alpha^5 . \tag{5.7}$$

Setting $\epsilon = 1$ does indeed give a reasonable estimate for diamond. Ordinary rocks have ϵ two or three orders of magnitude smaller.

Let us now apply these microscopic considerations to a macroscopic situation. Consider a planet of radius R and on it a mountain of height h. Suppose this mountain to be as high as it could be, without collapsing under its own weight. We can estimate h by equating the pressure at the bottom of the mountain to K, as

$$\rho g h \approx K, \tag{5.8}$$

where ρ is the density of the planet (and the mountain) and g is the surface gravity. Approximating $g \approx 4\rho R$ and then substituting for ρ and K from Eqs. (5.5) and

(5.7), respectively, gives $Rh \approx (4^2\epsilon/\alpha)(Am_e m_b)^{-2}$. The factor $4^2\epsilon/\alpha$ is not orders of magnitude from unity, and

$$Rh \approx \frac{1}{(Am_e m_b)^2} \tag{5.9}$$

is still a useful approximation. This expression predicts that smaller planets will have higher mountains, and sure enough, while Mars has only about half the radius of the Earth, Olympus Mons on Mars dwarfs any mountain on Earth. Continuing to smaller R, eventually h will be of the same order as R, and then what we have is no longer a mountain on a round planet, but an asteroid with an irregular shape. And indeed, rocky objects larger than a few hundred kilometres are round, whereas smaller objects have irregular shapes; in small objects interatomic forces are strong enough to maintain an irregular shape, whereas for large objects gravity is too strong and makes things round. Compare the asteroid Vesta (see APOD 110802) and the dwarf planet Ceres (APOD 160204). The formula (5.9) suggests that $1/(Am_e m_b)$ estimates the transition between an asteroid and a dwarf planet.

It is remarkable to see the microscopic quantity $1/(m_e m_b)$ taking on the role of a macroscopic length scale; we will see it again later in this chapter.

Exercise 5.1 Compute the following in SI units: (i) the density of ice according to equation (5.5), (ii) the bulk modulus for the hardest solids from formula (5.7) with $\epsilon = 1$ (diamonds have $K \simeq 500\,\mathrm{GPa}$), (iii) the maximum mountain height on Earth, according to formula (5.9).

We recommend writing short programs for such calculations, as the arithmetic becomes unwieldly, even on a calculator.

Exercise 5.2 As an example of intermolecular binding energies, consider the latent heat of melting for ice, which is $334\,\mathrm{kJ/\,kg}$. Calculate the effective ϵ such that the latent heat equals $\epsilon\, m_e \alpha^2$ per molecule.

Exercise 5.3 Estimate the pressure at the centre of a rock/ice planet in terms of its density and radius. If the Earth is approximated as having constant density, with $A = 30$, estimate the central pressure in SI units.

Exercise 5.4 Consider a solid planet with $A = 1$ (frozen hydrogen). For what M/M_L does the central pressure becomes comparable to non-relativistic degeneracy pressure (see Section 4.6)?

5.3 The Clayton Model

For anything more complicated than the constant-density case, a model in hydrostatic equilibrium typically requires numerical solution. There is one elegant example, though, due to D. Clayton, where $P(r)$ and $\rho(r)$ have simple analytical forms and yet the model remains a reasonably realistic description of stellar structure.

The model (Clayton, 1986) consists of a dense core of characteristic radius a, surrounded by a less-dense part to an outer radius R. The model has three parameters: the central density ρ_0, the radius R, and the core radius a. The total mass is a function of these three parameters, though of course the functional dependence can be inverted, if desired.

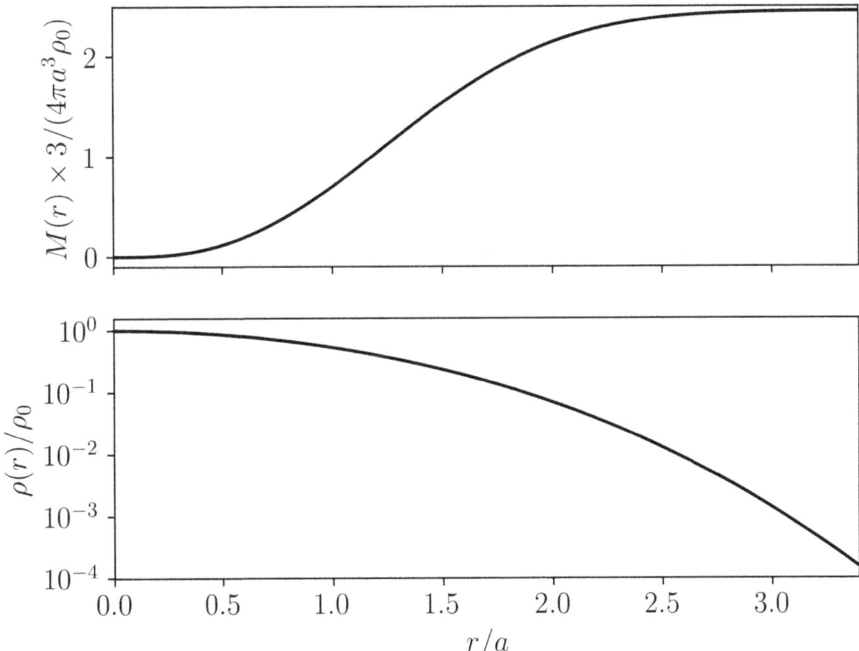

Fig. 5.2 Enclosed-mass and density profiles of the Clayton model. $M(r)$ in the upper panel is, apart from normalization, simply the function φ defined in Eq. (5.15). In the lower panel, note the logarithmic scale—$\rho(r)$ falls very steeply outside the core.

We start by writing down a plausible form for the pressure gradient[1]

$$\frac{dP(r)}{dr} = -\frac{4\pi}{3} G\rho_0^2 \, r \, e^{-r^2/a^2} \,, \tag{5.10}$$

which has the added convenience of being analytically integrable (indeed, by construction). For boundary conditions, we know that the centre will have the maximum pressure, and therefore the pressure gradient must go to zero at $r = 0$; additionally, the outer boundary is defined as the radius at which pressure goes to zero. Integrating (5.10) with these conditions in mind leads to the pressure profile:

$$P(r) = \frac{2\pi}{3} G\rho_0^2 \, a^2 \left(e^{-r^2/a^2} - e^{-R^2/a^2} \right) . \tag{5.11}$$

Starting from Eq. (5.10) again and combining it with hydrostatic-equilibrium from Eq. (5.3) leads to the relation

$$\frac{M(r)\,\rho(r)}{r^2} = \frac{4\pi}{3} \rho_0^2 \, r \, e^{-r^2/a^2} . \tag{5.12}$$

[1]For this section, we revert to non-Planckian units.

Multiplying each side of this equation with the corresponding side of mass-continuity in Eq. (5.4), and rearranging, yields

$$M^2(r) = \frac{32\pi^2}{3}\rho_0^2 \int_0^r r'^5 e^{-r'^2/a^2}\, dr' \,. \tag{5.13}$$

This integral can be done by parts. Doing so, and rearranging the result, gives

$$M(r) = \frac{4\pi a^3}{3}\rho_0\, \varphi(r/a)\,, \tag{5.14}$$

where

$$\begin{aligned} \varphi(x) &= \left[6 - 3(x^4 + 2x^2 + 2)e^{-x^2}\right]^{1/2} \\ &\approx \begin{cases} x^3 & \text{if } x \ll 1\,, \\ \sqrt{6} & \text{if } x \gg 1\,. \end{cases} \end{aligned} \tag{5.15}$$

Inserting $M(r)$ into Eq. (5.12) yields

$$\rho(r) = \rho_0\, \frac{r^3}{a^3}\, \frac{e^{-r^2/a^2}}{\varphi(r/a)}\,. \tag{5.16}$$

From the form of φ, we can see that this does satisfy our expectation that as $r \to 0$, then $\rho(r) \to \rho_0$, the central density. Figure 5.2 displays the form of $M(r)$ and $\rho(r)$.

With the density profile in hand, we can continue and derive the implied temperature profile. To do so, recall that stars are predominantly a plasma of protons (i.e., hydrogen) and electrons, which behave like a superposition of two classical monoatomic ideal gases. Additionally, to a very high degree, stars are charge neutral. With the number densities of protons and electrons being equal, $n_e = n_p = \rho/m_\text{b}$, the implied temperature profile temperature profile of a classical ideal gas (cf. Eq. (4.29)) could then be written

$$k_\text{B}T(r) = \frac{m_\text{b}}{2}\, \frac{P(r)}{\rho(r)}\,. \tag{5.17}$$

The Clayton model does not consider *how* the temperature profile is to be maintained—it merely describes a system that can be maintained in equilibrium with these properties. The thermal processes themselves come out of two competing processes: energy generation in the core, and energy transport through the rest of the star. Those processes are the subjects of the next two chapters.

5.4 The Virial Theorem (Again)

So far, hydrostatic equilibrium has been discussed in a local, shell-by-shell sense. Moving to a global view leads to a remarkable relation between gravitational and thermodynamic properties, leading to another form of the virial theorem (cf. Eq. (1.59)), this time in a continuous fluid context.

We start by looking back at hydrostatic equilibrium in Eq. (5.3), which describes a force balance between a pressure gradient and gravity. Let us multiply both sides of Eq. (5.3) by the volumetric term $4\pi r^3/3$ and integrate over r:

$$\int_0^R \frac{4\pi r^3}{3} \frac{dP(r)}{dr}\, dr = -\frac{1}{3} \int_0^R \frac{M(r)}{r}\, 4\pi r^2 \rho(r)\, dr. \tag{5.18}$$

The right-hand side has been written to make the factors in the integrand recognisable: we have the mass of an infinitesimal mass shell r, and multiplying it we have the gravitational potential energy per unit mass at r, due to the enclosed mass. The right-hand integral is thus the body's gravitational energy $E_{\rm grav}$. The left-hand side can be integrated by parts, and expressed as $-\langle P \rangle V$, where $\langle P \rangle$ means the volume-averaged pressure. Thus, a spherically symmetric object in hydrostatic equilibrium satisfies

$$\langle P \rangle V = -\tfrac{1}{3} E_{\rm grav}\,. \tag{5.19}$$

For the special case of constant density, the right-hand side of Eq. (5.18) can be worked out exactly, giving $M^2/(5R)$.

Let us now make a drastic approximation: let us neglect the radial dependence of both P and ρ entirely, focusing just on the scaling of the relationships. Eliminating V in the constant-density result in favour of R gives

$$P \approx \frac{M^2}{20R^4}\,. \tag{5.20}$$

Objects in hydrostatic equilibrium range over many order of magnitude in size, and as a result the internal variation of pressure and density in a single object is not as great as the variation between objects. For this reason the approximate virial theorem in Eq. (5.20) is extremely useful.

Exercise 5.5 Consider a classical gas star satisfying the usual relations in Eq. (4.29) and in hydrostatic equilibrium. Show that the kinetic energy of the particles equals $-E_{\rm grav}/2$.

5.5 Fermi-Gas Remnants I: Virial Approximation

Not everything with a star-like mass is a star. Some balls of gas have settled into hydrostatic equilibrium but lack the nuclear furnaces (described in Chapter 7) characteristic of real stars. Either their centres are not quite hot enough to start nuclear reactions, or they have exhausted their nuclear fuel. Without nuclear reactions heating up the gas (and maintaining a source of heat, as energy is radiated away), the gas pressure is not enough to stably oppose the gravity. Instead, a different source of outward-balancing force must be relied on, in this case from the pressure of a degenerate electron gas.

For hydrogen the number densities of baryons (in this case, just protons) and electrons are equal, so that again $n_e = n_p = \rho/m_{\rm b}$. Then, approximating the volume as $4R^3$ and recalling the mass scale in Eq. (5.1), one finds that

$$n_e \approx \frac{1}{4(m_{\rm b}R)^3} \frac{M}{M_{\rm L}}\,. \tag{5.21}$$

Inserting this number density into the (non-relativistic) degeneracy pressure from Eq. (4.44), and then substituting that pressure in the approximate virial theorem in Eq. (5.20) gives a mass–radius relationship for objects held stable by Fermi gas pressure:

$$R \approx \frac{3.5}{m_e m_b}\left(\frac{M}{M_L}\right)^{-1/3}. \tag{5.22}$$

This expression shows that the radius *decreases* with increasing mass: accreting mass would make the object shrink. Objects whose gravity is balanced by this electron pressure are called brown dwarfs. These objects tend to be dim (as their name suggests) and are therefore difficult to accurately count, but some examples can still be observed, such as APOD 951204, 991120, and 990324.

We can now use (5.22) to eliminate R from the number density (5.21). The corresponding Fermi level from Eq. (4.42) is

$$p_F \approx \frac{m_e}{2}\left(\frac{M}{M_L}\right)^{2/3}. \tag{5.23}$$

Recalling that $M_\odot \approx M_L/2$, we see that for an object of solar or sub-solar mass, $p_F \ll m_e$ and hence the electrons would indeed be non-relativistic. The most energetic particles would have energy ${p_F}^2/2m_e$ or

$$E_{\text{max}} \simeq \frac{m_e}{8}\left(\frac{M}{M_L}\right)^{4/3}. \tag{5.24}$$

If this energy were to become large enough for nuclear fusion processes to begin, we could get a real star (perhaps somewhat unfairly, brown dwarfs that *don't* acquire enough mass to ignite are sometimes referred to as 'failed stars'). The coldest observed brown dwarf to date has a surface temperature of approximately 27°C (APOD 110830)—it is amazing to think of being able to stand on the surface of such a quasi-star.

Another kind of non-star occurs when a star has converted all its hydrogen to helium, and the particles do not have enough energy to fuse into heavier elements. Nuclear reactions end, leaving a helium stellar remnant. Helium has half as many electrons ($2e^-$) as baryons ($2p^+ + 2n$), but the mechanism of electron degeneracy pressure maintaining the remnant is as before in the case of the brown dwarf. The relations from the previous section can be recycled, using the replacements

$$m_b \to 2m_b \qquad M_L \to M_L/4 \tag{5.25}$$

for the new case. This leads to a higher Fermi level than for the hydrogen remnant, and helium remnants are thus hotter than hydrogen remnants of the same mass. The former can also be more massive than the latter remnants, because massive hydrogen objects more quickly reach the ability to ignite nuclear reactions. As a result of these combined features, helium remnants tend to be typically much hotter than brown dwarfs, and they are known as white dwarfs. Many white dwarfs are observed cooling down from their previous nuclear-driven lives as stars, and are observed in clusters APOD 000910 and at the centres of (often extremely beautiful) nebulae, e.g., APOD 150517, 170419, and 121030.

The higher Fermi level also means that the electrons become relativistic for lower masses than is the case for hydrogen remnants. It is interesting to consider the extreme relativistic case, where the outward pressure is provided entirely by ultra-relativistic electrons. Using the number density from Eq. (5.21)—but halved for helium—in the extreme-relativistic pressure from Eq. (4.44) gives $P \approx 0.05\,(M/m_{\rm b})^{4/3}/R^4$. Substituting this pressure in the approximate virial theorem in Eq. (5.20) gives a solution for only one mass value:

$$M \approx M_{\rm L}\,. \tag{5.26}$$

This mass represents the upper limit for which electron degeneracy can still balance that mass's gravitational squeezing. It is a rough estimate of what is known as the Chandrasekhar limit, the maximum mass of a white dwarf. The precise value ($M_{\rm ch} = 0.78\,M_{\rm L}$) can be obtained by a more detailed calculation, outlined in Section 5.6 below.

For masses beyond the Chandrasekhar limit, electron degeneracy pressure is not able to hold up a helium remnant. It collapses into an even denser object until a new physical phenomenon can 'catch' it and provide a balancing force against gravity: in this case, the degenerate gas pressure of *neutrons* can provide the source. The results from the previous section can now be recycled again, this time replacing m_e with $m_{\rm b}$. Instead of the radius (5.22), now

$$R \approx \frac{3.5}{m_{\rm b}^2}\left(\frac{M}{M_{\rm L}}\right)^{-1/3} \tag{5.27}$$

applies. Such objects are known as neutron stars, though they are not really stars. They are about as exotic as you can get, this side of a black hole: imagine something the size of a small city, with as much mass as the Sun, and moreover spinning like a kitchen blender. The rapid spin arises from the angular momentum inherited from the progenitor star; more stars spin slowly, but a much smaller object with the same angular momentum must spin correspondingly faster. Some neutron stars produce a striking pulsed emission, because of their extreme spin and the magnetic fields around them. These are known as pulsars. Pulsars are a sub-field of astrophysics in their own right, partly because they are made up of an enigmatic state of matter (nuclear matter on a macroscopic scale), and partly because some pulsars are superb natural clocks. We will not delve into them in this book, but please admire the beautiful APOD 050326 of the Crab pulsar. The discovery of pulsars in the 1960s by Jocelyn Bell and colleagues is also a remarkable story (Byers and Williams, 2006).

Exercise 5.6 For neutron stars, since radius decreases as mass increases, some M will correspond to $R = 2M$. In other words, the remnant will become a black hole. Estimate the mass for which that happens. (It is known as the Tolman–Oppenheimer–Volkoff or TOV limit.) Will the degenerate neutron gas be relativistic or non-relativistic at the TOV limit?

5.6 Fermi-Gas Remnants II: Numerics

The drastic approximation of replacing the two differential equations for mass continuity and hydrostatic equilibrium with the simple virial relation (5.20) is useful, but

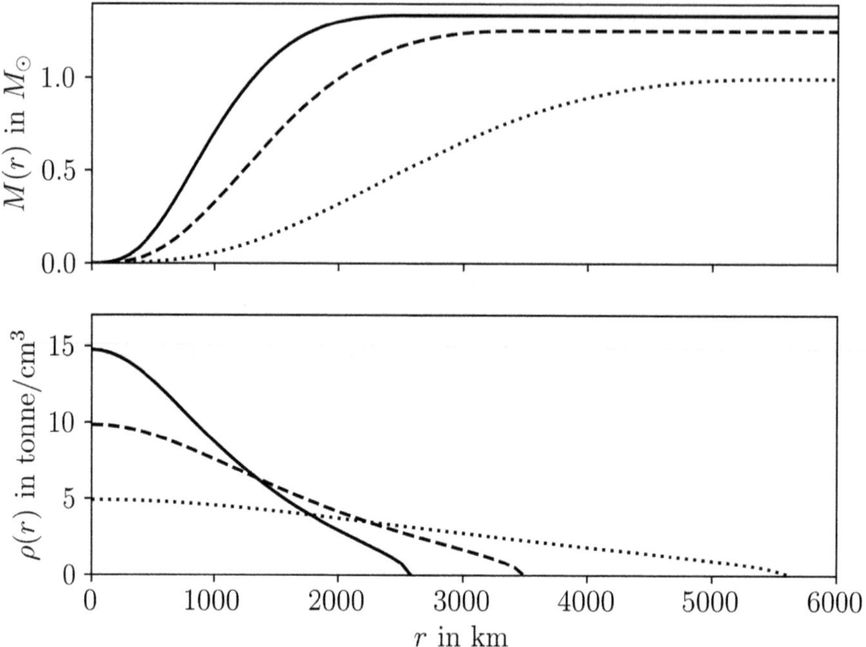

Fig. 5.3 Density and enclosed-mass profiles for three helium white dwarfs. The outer surface is the radius at which the density falls to zero and the enclosed mass levels off. Matched curves in the lower and upper panels have similar line style, from which we note that larger central density implies smaller radius and larger mass. Note the density units of tonne/cm^3 or 10^9 kg/m^3.

it is only a step towards better approximations. Let us now return to the differential equations (5.4) and (5.3) and see how to solve them for matter whose pressure comes from a degenerate electron gas that may or may not be relativistic.

A convenient simplification, though it works only for degenerate matter, is to change our variable of interest from pressure P to the Fermi momentum p_{F}. Differentiating the expression in Eq. (4.43) for electron degeneracy pressure, and keeping in mind the expression in Eq. (4.42) for electron density n_e, we have

$$\frac{dP}{dr} = n_e(p_{\mathrm{F}})\, v(p_{\mathrm{F}})\, \frac{dp_{\mathrm{F}}}{dr}\,. \tag{5.28}$$

For $v(p_{\mathrm{F}})$ the relativistic speed Eq. (4.9) is to be used. The mass density can also be written in terms of the Fermi momentum: writing μ for the number of baryons per electron ($\mu = 2$ for helium), the density can be expressed as

$$\rho = \mu\, m_{\mathrm{b}}\, n_e(p_{\mathrm{F}})\,. \tag{5.29}$$

Substituting in the preceding two expressions, the original equations (5.3) and (5.4) for gravity versus pressure become

$$\begin{aligned} \frac{dp_{\mathrm{F}}}{dr} &= -\frac{\mu\, m_{\mathrm{b}}\, M}{r^2\, v(p_{\mathrm{F}})}\,, \\ \frac{dM}{dr} &= 4\pi r^2 \mu\, m_{\mathrm{b}}\, n_e(p_{\mathrm{F}})\,. \end{aligned} \tag{5.30}$$

The functions $v(p_{\mathrm{F}})$ and $n_e(p_{\mathrm{F}})$ are given by Eqs. (4.9) and (4.42), respectively.

One can integrate Eqs. (5.30) outward from $r = 0$, with $M(0) = 0$ and some chosen central $p_{\mathrm{F}}(0)$ as initial conditions. When p_{F} falls to zero, the integration has reached the surface, determining the final radius R, and mass M. Figure 5.3 shows some examples, plotting the density ρ instead of p_{F} via their relation in Eq. (5.29).

On a practical note, the extreme exponents in Planckian units may cause numerical integration libraries to misbehave. Therefore, it is best to introduce rescaled variables, such as $\tilde{r}$, $\tilde{p}$, and $\tilde{M}$, defined as follows:

$$r \equiv \frac{\tilde{r}}{\mu m_e m_{\mathrm{b}}}\,, \qquad p_{\mathrm{F}} \equiv m_e \tilde{p}, \qquad M \equiv \frac{\tilde{M}}{\mu^2 m_{\mathrm{b}}^2}\,. \tag{5.31}$$

Changing to these scaled variables, the Eqs. (5.30) become

$$\begin{aligned} \frac{d\tilde{p}}{d\tilde{r}} &= -\sqrt{1 + 1/\tilde{p}^2}\,\frac{\tilde{M}}{\tilde{r}^2}\,, \\ \frac{d\tilde{M}}{d\tilde{r}} &= \frac{4}{3\pi}\tilde{r}^2\tilde{p}^3\,. \end{aligned} \tag{5.32}$$

Exercise 5.7 Integrate the Eqs. (5.32) numerically and plot $M(r)$ against r for different values of $p_{\mathrm{F}}(0)$. You can avoid the zero denominators at the centre and surface by adding a very small constants to $\tilde{r}^2$ and $\tilde{p}^2$. As $\tilde{p}(0)$, and hence the central density, is increased to very large values, the surface radius keeps getting smaller, while the total mass refuses to climb above the Chandrasekhar limit.

6

Nuclear Fusion in Stars

At the end of the 19th century, when astrophysics was still almost entirely about gravitational phenomena, gravity itself was also thought to be the primary energy source for stars as well. The reasoning was essentially as follows. The measured solar luminosity and mass translate into

$$L_\odot \simeq 0.7 \times 10^{-13} M_\odot/\mathrm{yr}\,. \tag{6.1}$$

Chemical combustion to H_2O or CO_2 releases $\simeq 10^{-10}$ of mass as energy. Hence, chemical burning could not even have kept the Sun shining through recorded history. On the other hand, the formation of a gas ball of solar mass and radius through gravitational infall of primordial gas would have released energy out of the gravitational potential. The current gravitational energy of the Sun is $E_{\mathrm{grav}} \sim M_\odot^2/R_\odot$ (cf. Eq. (5.18)). From the virial theorem, half of E_{grav} would have been radiated away. Noting that $M_\odot/R_\odot \simeq 2.5 \times 10^{-6}$ and comparing with Eq. (6.1) gives $\sim 10^7$ yr for the lifetime of the Sun. Even in 1900, however, there were indications that 10^7 yr is much too young, because geology and early evolutionary biology implied that the Earth is at least 10^9 yr old. Some new energy source was being hinted at.

A further hint came in 1905, with the association of rest mass directly with an energy, the famous $E = mc^2$. In his celebrated paper formulating the equivalence of mass and energy, Einstein suggested that atomic-weight measurements would be a good place to look for mass being converted to energy. By 1920, measurements of atomic weights indicated that a helium nucleus was roughly a percent less massive than four hydrogen nuclei. Eddington then suggested that nuclear fusion of hydrogen to helium was the energy source of the Sun. (He also prophetically wondered whether nuclear fusion might one day be used 'for the well-being of the human race—or for its suicide'.[1])

While Eddington was on the right track, it was not yet clear *how* nuclear reactions could actually take place in a star. Due to their constituents, nuclei have an inherently positive charge, and the electrostatic repulsion between them seemed an impossible obstacle to overcome to bring about nuclear fusion. At high-enough temperatures, the kinetic energy of the nuclei could theoretically overcome the electrostatic repulsion and allow the short-range nuclear force to perform fusion. But stars are not that hot. In fact, stars use a more subtle process: quantum tunnelling, as was first realized by George Gamow in the 1920s and worked out in detail by Hans Bethe in the 1930s.

[1] Address to the Mathematical and Physical Science Section of the British Association for the Advancement of Science, 1920.

The Astronomer's Magic Envelope. Prasenjit Saha and Paul A. Taylor.

10.1093/oso/9780198816461.001.0001

6.1 The Reactions

There are several possible fusion reaction paths from H to He. In the Sun the most common sequence, due to its mass and temperature, is the following (note that p is the same as ^{1}H):

$$\begin{aligned} p + p &\to {}^2\mathrm{H} + e^+ + \nu_e \\ {}^2\mathrm{H} + p &\to {}^3\mathrm{He} + \gamma \\ {}^3\mathrm{He} + {}^3\mathrm{He} &\to {}^4\mathrm{He} + 2p\,. \end{aligned} \tag{6.2}$$

The net result is that two sets of proton pairs combine to form a nucleus of ^{4}He, along with by-products: positrons (e^+), neutrinos (ν_e), and photons (γ) to balance charge, spin, and momentum. The total restmass of the products being less than that of the reactants, the difference goes into the γ and into the kinetic energy of the other products, both of which serve as heat sources for the hydrodynamic body. The sequence of reactions in Eq. (6.2), starting with the fusion of proton pairs to produce helium, is known as the *pp-chain.*

Another important set of nuclear reactions for converting hydrogyen to helium makes use of C, N, and O nuclei as catalysts. An example of such *CNO cycles* is the following:

$$\begin{aligned} {}^{12}\mathrm{C} + p &\to {}^{13}\mathrm{N} + \gamma \\ {}^{13}\mathrm{N} &\to {}^{13}\mathrm{C} + e^+ + \nu_e \\ {}^{13}\mathrm{C} + p &\to {}^{14}\mathrm{N} + \gamma \\ {}^{14}\mathrm{N} + p &\to {}^{15}\mathrm{O} + \gamma \\ {}^{15}\mathrm{O} &\to {}^{15}\mathrm{N} + e^+ + \nu_e \\ {}^{15}\mathrm{N} + p &\to {}^{12}\mathrm{C} + {}^4\mathrm{He}\,. \end{aligned} \tag{6.3}$$

CNO cycles generally require a higher starting energy than pp-chains and are therefore more relevant processes for larger stars, where they tend to dominate energy production.

Exercise 6.1 Tables of atomic weights give values of 1.0078 and 4.0026 for ^{1}H and ^{4}He, respectively (to four significant digits). As the reactions in Eqs. (6.2) and (6.3) indicate, nuclear fusion releases two neutrinos per ^{4}He atom output. Combine these two facts to estimate the number of neutrinos per joule generated by the Sun.

6.2 Quantum Tunnelling and the WKB Approximation

Stars with masses down to $\simeq 0.1 M_\odot$ (or $\simeq 0.05 M_\mathrm{L}$) are commonly observed. When such stars were in their proto-stellar stage, their most energetic particles would have had a kinetic energy of $\simeq 0.002\, m_e$ (see Eq. (5.24)). Such energies (of order 1 keV) are easy to achieve in laboratories here on Earth, and they do not yield nuclear fusion. To see the ignition problem in another way, let us compare the kinetic energy of two protons to their electrostatic potential energy α/r. Doing so gives the closest possible approach of $r = \alpha/0.002 m_e \simeq 4/m_e$. This distance (of order a picometre, 10^{-12} m) is orders of magnitude greater than the range over which the nuclear strong force can override the electrostatic force and make nuclear fusion possible (on the order of a femtometre, 10^{-15} m). Clearly, something else must be involved. That something, as we have already mentioned, is quantum tunnelling. It happens very, very rarely, but on the scale of the star, just often enough. To study the process of quantum tunnelling we will need the Schrödinger equation.

Let us first consider the classical dynamics of two nuclei, with A_1, A_2 and Z_1, Z_2 as their mass numbers and atomic numbers, respectively. Analogously to the gravitational two-body problem (cf. Eq. (1.42)), the dynamics can be described as an effective one-body problem with a reduced mass

$$M_{\rm red} = \frac{A_1 A_2}{A_1 + A_2} m_{\rm b} \,. \tag{6.4}$$

We make the further simplifying assumption to treat the dynamics as one dimensional, with r being the mutual distance. The Hamiltonian for this simple system is then

$$H(r,p) = \frac{p^2}{2M_{\rm red}} + V(r)\,, \tag{6.5}$$

where

$$V(r) = \frac{Z_1 Z_2\, \alpha}{r}\,. \tag{6.6}$$

In the corresponding quantum problem, the classical Hamiltonian, a function of position r and momentum p, is replaced by the Schrödinger equation for the wave function $\psi(r)$, yielding

$$E\,\psi(r) = -\frac{1}{2M_{\rm red}} \frac{d^2\psi(r)}{dr^2} + V(r)\,\psi(r)\,. \tag{6.7}$$

In this new quantum formulation, 'solving' the equation means determining $\psi(r)$.

We are particularly interested in describing the wave function in the region where $E < V(r)$. This corresponds to $r < r_{\rm t}$, where

$$r_{\rm t} \equiv \frac{Z_1 Z_2\, \alpha}{E} \tag{6.8}$$

is the radius at which the tunnelling commences, which depends on E. In this regime, $\psi(r)$ can be solved for by making use of the WKB approximation.[2] Let us write the wave function for tunneling from an outer radius $r_{\rm t}$ to some inner value in the form

$$\psi(r) = \exp\left(-\int_r^{r_{\rm t}} S(r')\,dr'\right)\,, \tag{6.9}$$

which is still quite general and now leaves us the function $S(r)$ to solve for. The Schrödinger equation in Eq. (6.7) then reduces conveniently (note that the exponential of S drops out of each term) to

[2]Named after Wentzel, Kramers, and Brillouin, three of its early users in quantum mechanics.

$$E = -\frac{1}{2M_{\rm red}}\left[S^2(r) - \frac{dS(r)}{dr}\right] + V(r)\,. \tag{6.10}$$

For the system at hand, the additional assumption is made that $S(r)$ is slowly varying, with the consequence that $dS(r)/dr$ is small compared to the other terms and can be dropped. Comparing with the Hamiltonian in Eq. (6.5), we see that S^2 behaves like minus the classical momentum squared. We now have a relatively simple expression for S:

$$S^2(r) \approx 2M_{\rm red}\left[V(r) - E\right]\,. \tag{6.11}$$

Substituting this into Eq. (6.9) and rearranging yields

$$\psi(r_0) \approx \exp\left(-\sqrt{2M_{\rm red}E}\int_{r_0}^{r_{\rm t}}\sqrt{\frac{r_{\rm t}}{r'} - 1}\,dr'\right)\,. \tag{6.12}$$

From this, the probability density τ for tunnelling through the potential to some radius r is calculated directly from the wave function:

$$\tau(r) = |\psi(r)|^2\,. \tag{6.13}$$

That is, $\tau(r)$ quantifies the transmission probability down to a certain depth through the repulsive potential. Of interest in this case is the probability of tunnelling down to essentially $r_0 \approx 0$ for some given E, so we denote this value in terms of energy as $\tau(r = 0) \equiv \tau(E)$. For $r_0 = 0$ the integral in Eq. (6.12) can be solved through the substitution $r' = r_{\rm t}\,\sin^2(u)$, after which it evaluates to $\pi r_{\rm t}/2$ and

$$\psi(0) \approx \exp\left(-\pi\, r_{\rm t}\sqrt{M_{\rm red}E/2}\right)\,. \tag{6.14}$$

We obtain the transmission probability from the wave function through the definition of τ in Eq. (6.13). We replace the tunneling radius with its constituents, since $r_{\rm t}$ depends explicitly on E:

$$\tau(E) \approx \exp\left(-\pi Z_1 Z_2\,\alpha\sqrt{2M_{\rm red}/E}\right) = \exp\left(-\sqrt{E_{\rm G}/E}\right)\,. \tag{6.15}$$

In characterizing the transmission probability with respect to E, we have defined the relevant energy scale $E_{\rm G}$ for the particle as

$$E_{\rm G} \equiv 2M_{\rm red}(\pi\alpha Z_1 Z_2)^2 = 2m_{\rm b}\frac{A_1 A_2}{A_1 + A_2}(\pi\alpha Z_1 Z_2)^2\,, \tag{6.16}$$

which is called the Gamow energy. It is interesting to note that Gamow also used this argument in the late 1920s to show a means for alpha particles to spontaneously tunnel *out* from the nuclear potential in radioactive decay.

Exercise 6.2 Consider an interaction between two hydrogen atoms with the electrostatic barrier shielded outside $r_{\rm B} = 1/(m_e\alpha)$. In effect, the integral in Eq. (6.12) is $\int_0^{r_{\rm B}}$. Show that if $E \ll m_e\alpha^2$ the tunnelling probability is

$$\tau = \exp(-4\sqrt{m_{\rm b}/m_e})\,,$$

independent of E. For hydrogen atoms the probability is too small to be interesting. But if the electrons are replaced by muons (200 times heavier), muon-catalyzed fusion occurs.

6.3 The Reaction Rate

For a process involving two reactants A and B, the *reaction rate* is the number of reactions per unit volume per unit time, which can be written as

$$n_A n_B \langle \sigma v \rangle \,. \tag{6.17}$$

Here n_A and n_B are the respective number densities of each reactant, and intuitively it seems appropriate that for higher densities the rate should increase (as there should be a higher probability for encounters between the reactants). The factor $\sigma = \sigma(v)$ denotes the probability that an encounter at some relative velocity v will result in a reaction. For consistency, σ must have dimensions of area, and this quantity is called the reaction *cross section*, which will be analysed further later. These definitions and description apply generally to both chemical and nuclear reactions. Since the rate of encounters occurring is naturally proportional to the relative velocities, $\sigma(v)$ is weighted by v and then averaged over the distribution of relative velocities. Writing $P(v)$ for the the system's probability distribution of particle speeds, we have

$$\langle \sigma v \rangle = \int_0^\infty \sigma(v)\, v\, P(v)\, dv \,. \tag{6.18}$$

To make use of the rate description in Eqs. (6.17) and (6.18), one must specify the expected probability distributions $P(v)$ of the system as well as the cross sections $\sigma(v)$ for the reactants involved. For the case of stellar interiors, we first assume that we are studying a Maxwell–Boltzmann gas, so that $P(v)$ is given from Eqs. (4.25) and (4.31) and is simply a 3D Gaussian. We now work to estimate the σ for the reaction rate. Since we are dealing with a kinetic system, we can equivalently discuss velocity or energy and convert $\sigma(v) \to \sigma(E)$ directly with the classical definition of kinetic energy $E = M_{\text{red}} v^2/2$. As for the cross section's form, we can think of it in terms of three probabilistic components, each of which depends only on kinetic energy and which get multiplied together to form the total:

$$\sigma(E) = A(E)\, \tau(E)\, S(E) \,. \tag{6.19}$$

Here, $A(E)$ is the probability of the atoms running into each other; $\tau(E)$ is the probability of the atoms overcoming a repulsive barrier (i.e., transmission probability); and $S(E)$ is the probability of interacting/fusing once through the barrier.

A full understanding of σ requires a good deal of nuclear theory and/or accelerator experiments. Here we will simply note that $A(E)$ is like an area—the effective cross section each reactant particle presents, in being able to interact meaningfully with another. According to the de Broglie relation (Eq. (4.37), in which length is inversely related to momentum p), the cross section area will be $\approx 1/p^2$. Assuming the particles are non-relativistic, $1/p^2 \propto 1/E$. Accordingly, we approximate that $A(E) \propto 1/E$. The $\tau(E)$ factor we have derived earlier has the form of Eq. (6.15). The remaining cross-sectional factor in Eq. (6.19) encompasses the complex realm of the strong nuclear force, but to a first approximation, the probability of interaction is roughly constant across the energies we investigate here: $S(E) \approx S_0$.

This all leads to an energy-dependent expression for the cross section:

$$\sigma(E) = \frac{S_0}{E} \exp\left(-\sqrt{E_G/E}\right) . \tag{6.20}$$

Incorporating the Maxwell–Boltzmann velocity distribution and changing variable from v to E, one obtains

$$\begin{aligned} \langle \sigma v \rangle &= \int_0^\infty \sigma(E)\, v \, \frac{e^{-E/T}}{(2\pi T/M_{\text{red}})^{3/2}} \, 4\pi v^2 \, dv \\ &= \frac{4 S_0}{\sqrt{2\pi M_{\text{red}} T}} \int_0^\infty e^{-\sqrt{E_G/E}} \, e^{-E/T} \, dE/T \, . \end{aligned} \tag{6.21}$$

The integrand has an interesting form for both physical interpretation and for mathematical approximation. The first exponential factor is known as the Gamow factor, and it increases with energy. The accompanying exponential is called the Boltzmann factor, and it *decreases* with increasing energy. Due to the multiplication of such competing dependencies, the integrand as a whole peaks at some intermediate energy (labelled as the Gamow peak). The (relatively narrow) range of energies around this peak before either term dominates to reduce the integral essentially zero is called the Gamow window. It is in this band of kinetic energies that nearly all fusion reactions happen, and there is negligible contribution to the integral from outside this window.

When calculating the Gamow window and then solving the integral (6.21), the introduction of the dimensionless number

$$\Theta \equiv \left(\frac{E_G}{4T}\right)^{1/3} \tag{6.22}$$

helps simplify the algebra. The definition of Θ is chosen such that the exponent in (6.21) peaks at $E = \Theta T$. Further, when the dimensionless variable $u = E/T$, the integral part of (6.21) becomes

$$\int_0^\infty \exp\left(-2\Theta\sqrt{\Theta/u} - u\right) \, du \, . \tag{6.23}$$

One can check that the exponent (again, coming from the competing Gamow and Boltzmann factors) has a maximum at $u = \Theta$ (and please do!). Taylor-expanding around the maximum, we have

$$-2\Theta\sqrt{\Theta/u} - u \approx -3\Theta - \frac{3}{4\Theta}(u - \Theta)^2 \, , \tag{6.24}$$

and this provides an approximation that helps in simplifying the integral in (6.23) to be

$$\exp(-3\Theta) \int_0^\infty \exp\left(-\frac{3}{4\Theta}(u - \Theta)^2\right) \, du \, . \tag{6.25}$$

This technique of approximating an integral in this way is often called the Laplace approximation; see Fig. 6.1 for a graphical representation. The integral in Eq. (6.25)

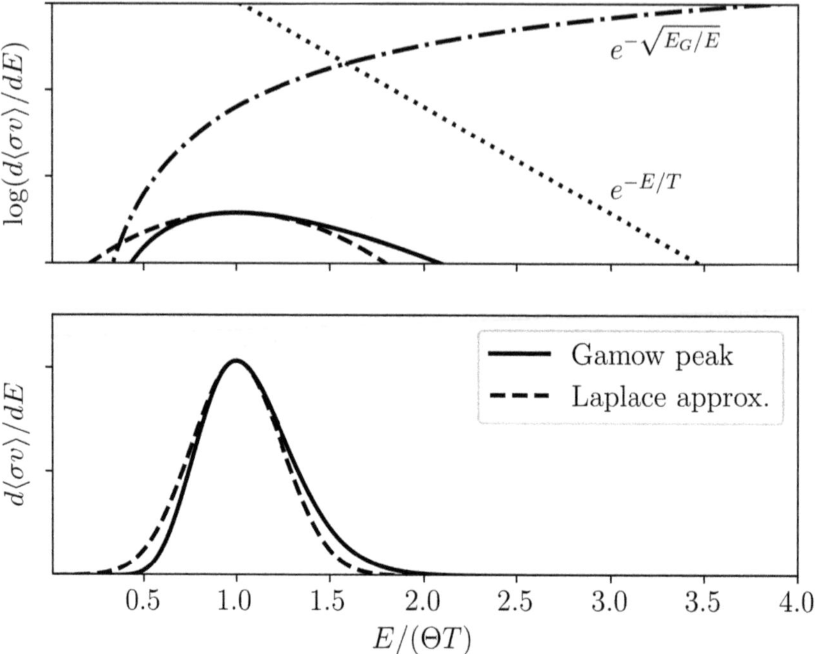

Fig. 6.1 The top panel shows the individual Gamow and Boltzmann factors (in log scale), and their combined contribution to forming the Gamow peak around $E = \Theta T$. Also shown is the approximate Gaussian function approximating the overall curve, with the goodness of fit particularly demonstrated in the lower panel's linear graph.

is a Gaussian integral with value $\sqrt{\pi\Theta/3}$. Inserting the result into the earlier expression (6.21) gives us

$$\langle \sigma v \rangle \approx 4S_0 \left(\frac{\Theta}{6M_{\mathrm{red}}T} \right)^{1/2} \exp(-3\Theta\,) \,. \tag{6.26}$$

The remaining parameter of interest in the reaction rate is the temperature. Recall that we had initially started down this path of derivations by asking at which temperature (or mass of proto-star) fusion reactions would begin in earnest. The overall T-dependence in Eq. (6.26) has the rather curious form $\langle \sigma v \rangle \propto x^{-2} e^{-1/x}$, where $x \propto T^{1/3}$ (recall that Θ depends on T). The reaction rate is highest for $T \simeq E_{\mathrm{G}}$, falling steeply at lower temperatures and slowly at higher temperatures. The low-T drop-off is, however, not as steep as an exponential. Thus, fusion can start and be sustained even for $T \ll E_{\mathrm{G}}$. Physically, what has happened is that nuclei in the high-E tail of a Maxwell–Boltzmann gas may achieve fusion through quantum tunnelling even when the distribution's peak energy (or velocity) is much lower than the Gamow energy.

It is also of interest to take further note of the Gamow energy formulation in Eq. (6.16). E_{G}, and hence Θ, increases with both atomic and mass numbers. As a result, higher Z and A give a steeper T-dependence. In particular, since CNO cycles involve heavier nuclei than pp-chains, they will have steeper temperature dependence. Thus, CNO cycles become progressively more important as temperature increases, becoming dominant in more massive stars.

Stars also have a simple feedback loop which regulates the reaction rate: if the temperature rises, the density falls (by physical expansion), and vice versa. Thus, they have a certain stability, which we on Earth, revolving around the Sun for millions of years, might consider a useful property.

Exercise 6.3 Of the reactions in Eq. (6.2), which one depends most strongly on temperature? Which one depends least strongly?

Exercise 6.4 The temperature dependence of the reaction rate is sometimes represented by an effective exponent

$$\langle \sigma v \rangle \propto T^{\beta}\,, \qquad \beta = \frac{d\ln\langle \sigma v \rangle}{d\ln T}\,.$$

Show that $\beta = \Theta - 2/3$.

7 The Main Sequence of Stars

At the end of the previous chapter we noted that stars have a feedback loop which keeps them shining steadily. In this chapter we will study the main process underlying that feedback loop: radiative transfer. The heat created by nuclear reactions in the star's interior leaks outward, behaving like a photon gas slowly diffusing out to larger radii. In the process, the radiation 'gas' deposits energy into layers of the star, heating up the particles and providing a pressure source to maintain a balance against gravity. Gravity itself serves to squeeze a star, leading to greater pressure and more nuclear reactions in the interior; the produced electromagnetic radiation heats up outer layers causing them to expand against gravity, leading to lower internal pressure and subsequent overall cooling and therefore slight compression, and onward in a generally balanced cycle.

Radiative transfer is not the only energy-transport process in stars. Convection also plays a role, especially in low-mass stars, transferring heat from the center through interchange of heated material to cooler layers, with outer material sinking to the interior. In very special circumstances, neutrinos play an important role in energy transport with dramatic results. Whereas convection and radiation processes transfer central heating into outer layers, neutrinos travel directly out of the star instead of diffusing, due to their neutral charge and extremely small cross sections of interaction. This extreme energy leakage breaks the feedback cycle regulating the star and can indirectly cause it to overheat, self-destructing as a supernova.

The study of radiation goes back to ancient times (think of burning glasses and mirrors) but in the astrophysical context radiation theory dates from the early 20th century. Karl Schwarzschild considered radiative transfer between parallel planes to model a stellar atmosphere, and then Eddington extended the ideas to spherical geometry for stellar interiors. In the next generation, important contributors included Fred Hoyle, Martin Schwarzschild, and the prolific Chandrasekhar.

7.1 Opacity and Radiative Transfer

Inside a star, a photon can travel only a short distance before being *scattered* by an electron—that is, it gets redirected or absorbed and re-emitted, possibly at a different energy. In macroscopic terms, part of the radiation flux at any location is absorbed and a fraction re-emitted (essentially isotropically). This combination of phenomena is known as opacity.

The Astronomer's Magic Envelope. Prasenjit Saha and Paul A. Taylor.

10.1093/oso/9780198816461.001.0001

Opacity can be characterized by assigning each electron an effective area or cross section σ for scattering photons.[1] One can imagine each electron as a deflecting lens, through which refraction and redirection occur. The opacity is often defined as a quantity per unit mass κ (thus, for hydrogen, $\kappa = \sigma/m_{\mathrm{b}}$). While the latter usage (with κ) tends to be more common in calculations, we will work in terms of the per-particle opacity σ here, as it is more intuitive. In accordance with its geometrical association, the cross section is typically related to the length scale of the classical radius of the electron,

$$\begin{aligned} r_e &\equiv \frac{1}{4\pi\epsilon_0}\frac{e^2}{m_e c^2} = \frac{\alpha\hbar}{m_e c} \qquad &\text{[SI units]}, \\ r_e &\equiv \frac{\alpha}{m_e} \qquad &\text{[Planckian units]}. \end{aligned} \tag{7.1}$$

As an area through which interactions occur, the cross section is then described as being of order of the square of the classical radius of the electron,

$$\sigma \sim (\alpha/m_e)^2 . \tag{7.2}$$

The precise value of σ depends, however, on the local composition, density and temperature. Determining σ accurately from atomic physics is, in practice, one of the most complicated aspects of stellar structure.

Adding to the stellar structure equations in Chapter 5, the fundamental equation in this chapter is the radiative transfer equation in spherical geometry,

$$\frac{dP_\gamma(r)}{dr} = -\sigma n_e \frac{L(r)}{4\pi r^2}, \tag{7.3}$$

where $P_\gamma(r)$ is the radiation pressure at radius r. Using the photon gas relation in Eq. (4.35) that $P_\gamma = (\pi^2/45)T^4$, the derivative on the left-hand side of (7.3) can be expressed as a temperature derivative, thus providing the temperature equation for stellar structure promised in Chapter 5. The right-hand side of Eq. (7.3) involves the luminosity $L(r)$, which is the total outward rate of change of energy through a shell at radius r. In a steady state, that quantity describes the net energy production interior to r. The energy flux at r will be $L(r)/(4\pi r^2)$. Since $c = 1$ in our Planckian units, we can also think of the energy flux expression as describing the photon momentum per unit area per unit time (recall that $E_\gamma = p_\gamma c$, as there is no rest mass), which is to say, a pressure. Within a differential shell of thickness dr, a fraction $\sigma n_e\, dr$ of radiation is absorbed and re-emitted isotropically. The product equals the resulting change in radiation pressure dP_γ, and hence the relation in Eq. (7.3).

Exercise 7.1 Photons inside a star effectively travel in a random walk, with the step-length being the typical 'mean free path' between scatterings

$$l \sim \frac{1}{\sigma n_e} .$$

After s steps of such a random walk, the typical (root-mean-squared) linear distance travelled is $\sqrt{s}\, l$.

[1]Note that σ in this chapter is distinct from the specific nuclear-reaction cross section in the previous chapter.

Assuming that a star has $M \simeq M_{\rm L}$ and has approximately the density of water, show that the time scale for a photon to travel from the centre of a star to the surface is

$$\frac{\alpha^3}{m_e m_{\rm b}^2}.$$

How long is this in common units?

7.2 Luminosity and Effective Temperature

In Chapter 5 we used the virial theorem in the approximate form of Eq. (5.20) as a surrogate for the solution of the equation of hydrostatic equilibrium. In this chapter we adopt a similarly drastic approximation, by again ignoring radial dependence and investigating Eq. (7.3) through its scaling relationships:

$$\frac{P_\gamma}{R} \approx -\sigma n_e \frac{L}{4\pi R^2} \quad \rightarrow \quad \frac{L}{4\pi} \approx \frac{R P_\gamma}{\sigma n_e}. \tag{7.4}$$

This provides a scale-order surrogate for the radiative transfer equation in Eq. (7.3) where, as before, R is the stellar radius.

We now combine this radiative transfer information with the earlier hydrostatic equilibrium derivations. We use the virial relation in Eq. (5.20) as well as the fact that $n_e = \rho/m_{\rm b}$. Importantly, we note that that virial relation was also derived with scaling, and, taking the average density of the spherical star to be $\rho \approx M/(4R^3)$, the level of approximation remains consistent across the various relations. Combining these expressions with Eq. (7.4) and eliminating both R and n_e, we obtain the following relation of a star's luminosity to its mass:

$$L \approx 3 \times \frac{m_{\rm b}}{\sigma}\frac{P_\gamma}{P} M = \frac{3}{\sigma m_{\rm b}}\frac{P_\gamma}{P}\frac{M}{M_{\rm L}}. \tag{7.5}$$

Note that this expression involves both the total pressure P, which includes the gas pressure $P_{\rm gas}$ and the radiation pressure P_γ, and P_γ on its own.

Another quantity of interest is the effective temperature $T_{\rm eff}$, which is the temperature of a blackbody with the same luminosity and surface area. Harkening back to Eq. (4.35) again, the radiant intensity expression implies that

$$\frac{L}{4\pi R^2} = \frac{\pi^2}{60} T_{\rm eff}^4 \tag{7.6}$$

as the definition of $T_{\rm eff}$ in the present approximation. Substituting Eq. (7.5) into this equation and using the virial relation in Eq. (5.20) to eliminate R as before gives

$$T_{\rm eff}^4 \approx 6 \times \frac{m_{\rm b}}{\sigma}\frac{P_\gamma}{\sqrt{P}}. \tag{7.7}$$

In terms of combining theory with observation, the luminosity and effective temperature of a star are relatively straightforward quantities to measure: L requires

knowing the distance and observed brightness, while $T_{\rm eff}$ is given by the wavelength at which the star is brightest through Wien's displacement law in Eq. (4.40). A figure showing luminosity versus effective temperature for a set of stellar objects is known as a Hertzsprung–Russell diagram. An example appears in APOD 010223.

In practice, Hertzsprung–Russell diagrams do not often show $T_{\rm eff}$ and L explicitly. In place of these, usually *stellar magnitudes* are given. The magnitude scale for stars is basically a logarithmic luminosity scale; it relates the brightness of stars, which can vary over orders of magnitude, to that of the Sun $L_\odot$. It will not be used in this book, but we explain it briefly here. A star's luminosity L is related to its magnitude 'mag' as

$$L \approx 10^{-\frac{2}{5}\text{mag}} \times 10^2\, L_\odot\,. \tag{7.8}$$

That is to say, magnitudes are a logarithmic brightness scale such that $-2.5\,\text{mag}$ corresponds to a tenfold brightness increase, normalized such that zero magnitude corresponds to approximately $10^2\, L_\odot$. Thus, the Sun has mag $\simeq 5$, a star with mag $= 0$ is $\simeq 100$ times as luminous, and negative-magnitude stars still more luminous. This curious definition originates in the historical origin of the scale to classify stars in the night sky.

Note also that the relation in Eq. (7.8) is only an approximate relation, and not the formal definition. In detail magnitudes are defined more specifically in terms of power observed in different spectral intervals (passbands). There are different stellar magnitudes, known as U, B, V, R, or I, depending on whether the spectral region being observed is ultraviolet, blue, in the middle of the 'v'isible spectrum, red or infrared, respectively. In particular, V-band magnitude is

$$\text{mag}_V = 4.8 - \frac{5}{2}\log_{10}\left(\frac{L_V}{L_{\odot\,V}}\right). \tag{7.9}$$

(The approximate definition (7.8) above would give 5 as the constant in front, rather than 4.8 as here.) Since magnitudes measure brightness in a minus-logarithmic fashion, a magnitude difference like $B-R$ is the *ratio* of red to blue luminosity. Large $B-R$ implies comparatively more red light and hence a lower temperature. Magnitude differences such as $B-R$ or $B-V$ are known as colours and are used as a surrogate for $T_{\rm eff}$. Thus, another name for a Hertzsprung–Russell diagram is a colour-magnitude diagram, and this is also reflected in the axis labels of APOD 010223.

Exercise 7.2 As stars get older, they tend to get brighter. Why is that? *Hint:* What happens to n_e as a star ages?

7.3 High-Mass Stars

The most massive stars (say $10M_\odot$ or more) are, in some ways, the simplest to understand. We know from the virial theorem that higher mass leads to higher temperatures, and having very high temperatures simplifies two things. First, the pressure is predominantly due to radiation, so $P \approx P_\gamma$. Second, the opacity is dominated by non-relativistic scattering of photons by electrons, with nuclei playing no significant role; this is the regime of *Thomson scattering*, with cross section described by:

$$\sigma_{\mathrm{T}} = \frac{8\pi}{3}\left(\frac{\alpha}{m_e}\right)^2 . \tag{7.10}$$

Putting these factors into the expression (7.5) for the luminosity gives

$$L \approx \frac{1}{3}\frac{m_e^2}{\alpha^2 m_{\mathrm{b}}}\frac{M}{M_{\mathrm{L}}} . \tag{7.11}$$

Luminosities of stars are often quoted relative to the solar luminosity $L_\odot$, so it is useful to have the $L_\odot$ in terms of microscopic quantities. To do this, we can estimate the solar luminosity from the solar-energy flux on Earth (cf. Exercise 4.7) and the distance to the Sun—or simply look up the value needed—and compare with $m_e^2/(\alpha^2 m_{\mathrm{b}})$ expressed in the same units. The result is

$$L_\odot \simeq 2.5 \times 10^{-5}\,\frac{m_e^2}{\alpha^2 m_{\mathrm{b}}} . \tag{7.12}$$

As we can see, the formula in Eq. (7.11) would vastly over-predict the luminosity of the Sun. But the Sun is not a high-mass star and is not dominated by radiation pressure. A more appropriate comparison would be with the star Zeta Ophiuchi (ζ Oph), which has $M \simeq 10 M_{\mathrm{L}}$ and luminosity $\simeq 10^5 L_\odot$. Plugging these numbers into Eqs. (7.11) and (7.12), the predicted luminosity is quite respectable.

Any star has a finite supply of energy, which is essentially determined by its mass. The luminosity determines how quickly its fuel is consumed and radiated away, and so the lifetime of a star is roughly proportional to the amount of energy present divided by the luminosity. We know from atomic weight values (as discussed in Chapter 6) that the fusion of hydrogen to helium releases $\simeq 10^{-2}$ or 1% of the particle mass as photons. Looking at the luminosity–mass relation for massive stars in Eq. (7.11) and recalling that $M_{\mathrm{L}} = 1/m_{\mathrm{b}}^2$, then

$$\begin{aligned}\text{massive star's lifetime} &\simeq \frac{0.01M}{L}\\ &\simeq 3\times 10^{-2} \times \frac{\alpha^2}{m_e^2 m_{\mathrm{b}}} ,\end{aligned} \tag{7.13}$$

or a few tens of million years. Note that the lifetime is independent of M in this range of masses (at the present level of the scaling approximation).

Proceeding now to the effective temperature expression in Eq. (7.7) and putting in the same assumptions that P is entirely radiation pressure and σ is the Thomson cross-section, we have have the following relation for massive stars:

$$T_{\mathrm{eff}}^4 \approx 0.3\,\frac{m_e^2 m_{\mathrm{b}}}{\alpha^2}\,T^2 . \tag{7.14}$$

We can rewrite the last equation, interpreting the coefficient on the right as a temperature squared, as $T_{\mathrm{eff}}^2 \approx (130\,\mathrm{K}) \times T$. This relation indicates that the surface of a star is much less hot than its interior.

Exercise 7.3 Show that the estimate in Eq. (7.12) can be improved upon by combining the equations of hydrostatic equilibrium and radiative transfer together with Thomson scattering. The result

$$L(r) = \frac{3}{2}\frac{m_e^2 m_b}{\alpha^2} M(r)$$

is known as the Eddington luminosity. Express it in watts per kilogram.

7.4 Medium and Low-Mass Stars

In the stellar mass regime explored in the previous section, we were able to ignore the pressure due to the gas. One might subsequently ask for what masses the *reverse* case would be true—that is, when is the photon pressure negligible and $P \approx P_{gas}$ a reasonable representation of the system? Assuming a classical ideal gas, as described in Eq. (4.29) for ionized hydrogen, with protons and electrons contributing equally to the pressure, we have

$$P_{gas} = 2n_e T \,. \tag{7.15}$$

Writing n_e in terms of M, m_b, and R, and then again using the approximate virial theorem in Eq. (5.20) to eliminate R gives

$$T \approx \tfrac{1}{4}P^{1/4}\left(\frac{M}{M_L}\right)^{1/2} , \tag{7.16}$$

and hence

$$\frac{P_{gas}}{P_\gamma} \approx 10^3 \times \left(\frac{M}{M_L}\right)^{-2} \qquad (\text{for } P_{gas} \gg P_\gamma). \tag{7.17}$$

Clearly, for lower-mass stars, gas pressure will dominate.

Substituting the ratio from Eq. (7.17) into the expressions in Eq. (7.5) for the luminosity and in Eq. (7.7) for the effective temperature, and assuming that the opacity is Thomson, gives

$$L \approx 4 \times 10^{-4} \times \frac{m_e^2}{\alpha^2 m_b}\left(\frac{M}{M_L}\right)^3 \tag{7.18}$$

and

$$T_{eff}^4 \approx 10^{-2} \times \frac{m_e^2 m_b}{\alpha^2} T^2 \frac{M}{M_L} \,. \tag{7.19}$$

Recalling that $M_\odot = 0.55M_L$, we see that the luminosity expression in Eq. (7.18) gets about ~ 2.5 times the correct value of the solar luminosity in Eq. (7.12). In view of our crude approximation in Eq. (7.4) to the radiative transfer equation, the result is not too bad. The mass-dependence of the luminosity is also not bad: stars in the solar-mass range are observed to have $L \propto M^{3.5}$.

For lower-mass stars, the dependence of luminosity on mass gets steeper. The reason is that for lower temperatures, Thomson scattering by free electrons is no longer the main source of opacity. While the gas remains ionized, the electric fields of the nuclei play a significant role. For the lowest-mass stars, convection of the gas also becomes important for transporting energy. We will not pursue these regimes in this book.

Exercise 7.4 We can infer T_{eff} for the sun using the observation that $\lambda_W = 0.5\,\mu m$. We also know that the Sun has $M \simeq 0.5 M_{\text{L}}$. Using these values, compute what the estimate in Eq. (7.19) gives for T. Look up the solar interior temperature and comment.

8
The Expanding Universe

In the last part of this book, we turn our attention to cosmology, which is the study of the largest scales of the Universe, from birth until the present. APOD 120312 provides an enjoyable (and interactive) comparison of the scales involved.

Our present-day picture of the expanding Universe began to form in the early 20th century, around the same time as the mysteries of stellar structure began to be unravelled. The arrival of general relativity was for cosmology something like what quantum mechanics was for the understanding of stars: a radical new theory that raised all sorts of non-classical possibilities: what does space look like on (very) large scales? Does it curve? Are stars powered by nuclear reactions? What is the role of degeneracy pressure? But whereas for stars, plenty of data were already available and ready to confront any theoretical model, there were not really any such tests in cosmology. So it is not so surprising that, while an understanding of stellar structure developed quite quickly, ideas in cosmology would remain controversial for a long time. Among the early theoretical researchers in relativistic cosmology, Alexander Friedmann and Georges Lemaître are especially admired today. Somehow they managed to develop the basic theory of an expanding Universe back in the 1920s, when it was uncertain whether there even *was* anything beyond the Milky Way. But it was the subsequent observational revolution, and especially the work of Edwin Hubble, that really made the expanding Universe a successful paradigm.

8.1 On Measuring Distances

We noted at the beginning of Chapter 1 that the invention of telescopes around 1610 sparked the transformation of classical astronomy into astrophysics. Today, modern telescopes are a major and diverse subject of study in their own right. Telescopes are not only fascinating in how they capture and measure light, radio, and now even gravitational waves; they also continue to direct the development of astrophysics, even as they did in the 17th century. This book doesn't delve much into the observational realm of astrophysics, but there is one aspect of it that we need to touch upon now—namely, the role of distance measurement in cosmology.

Perhaps the most 'classical' method of measuring distances is parallax, which is (as mentioned in Chapter 1) the astronomical version of stereoscopic vision. From this technique comes the standard unit of distance in astronomy: a parsec (from *par*allax *sec*ond), defined as the distance at which 1 astronomical unit (au) subtends an angle of 1 arc-second. Recalling the definition of the au in Eq. (1.15), we get

$$1\,\mathrm{pc} = 1.029 \times 10^{8}\,\text{light-sec}\,. \tag{8.1}$$

The Astronomer's Magic Envelope. Prasenjit Saha and Paul A. Taylor.

10.1093/oso/9780198816461.001.0001

In popular astronomy, a parsec is commonly quoted as 3.26 ly, but 10^8 light-sec may be more convenient. For larger distances, kiloparsecs (kpc) and megaparsecs (Mpc) are used. Parallax is excellent for stars in our part of the Milky Way, and the GAIA mission is working on extending parallax measures to most of our Galaxy (see APOD `160926` and APOD `170417`). But that is less than 10% of the distance to the nearest large galaxy, Andromeda. Some variants of the parallax method use a moving *source* rather than a moving observer (while still making use of basic geometric relations of triangles). The most far-reaching applications of these parallax methods, also mentioned in Chapter 1, are provided by giant natural masers in distant galaxies. Masers out to scales of about 100 Mpc have now had their distances measured, providing useful beacons for pinpointing their host galaxies. But only a small fraction of galaxies have such so-called megamasers, so this technique is applicable to a comparatively small part of our extended 'neighbourhood'.

Distance measurements on cosmological scales must ultimately rely on very different strategies. A *standard candle* is the technical (and slightly whimsical) term for an object whose intrinsic luminosity L_{in} can be determined without also measuring its distance D, through some combination of theory and local observations. Measuring the apparent brightness L_{app} of a standard candle then gives a formulation for the distance via the geometry of spherical surfaces: $L_{\text{in}}/L_{\text{app}} \propto D^2$.

The first standard candles to be identified, and which are still very important today, were the *Cepheids*. These are stars whose outer regions have a pulsating mode due to a particular balance of radiative transfer and convection, and as a result their luminosity oscillates with a period of days or weeks. Importantly, by understanding the physics of the oscillation, their *intrinsic* luminosity can be inferred from the period. Henrietta Leavitt's discovery in 1912 of the period–luminosity relation for Cepheids was a watershed event in astronomy, because it opened the way to distance measurements outside our Galaxy. Nowadays Cepheid distances can be measured to $\sim$20 Mpc. (The early years of Cepheid distances are reminiscent of Huygens' measurement of the distance to Sirius by comparing its brightness to the Sun. Huygens measured the apparent brightness accurately, but its intrinsic brightness could only be guessed initially.)

Today, the most far-reaching distance measurements, however, come from an exceptionally bright standard candle, supernovae of the so-called Type Ia. These events come about through a special combination of circumstances, which are still not completely understood, but which can be outlined as follows. The progenitor star is a white dwarf in a binary-star system. If the binary is very close, the companion star may overflow its Roche lobe, and, as we discussed in Chapter 2, a runaway mass transfer to the white dwarf may occur. If the accretion happens to send the white dwarf mass over the Chandrasekhar limit, then the white dwarf's equilibrium cannot be sustained. The immediate result is collapse, but this collapse drives the temperature high enough that fusion reactions beyond helium can be ignited, producing a large amount of energy that results in an explosion. Importantly for us here, the physics of this particular explosion process yields fairly 'standard' (or, at least, standardizable) properties, assuming things like a fairly uniform critical mass limit and total combustion.

It should be noted that there are also other types of supernovae, arising mainly from massive collapsing stars rather than accretion onto white dwarfs. These events are *not* standard in the same way, and must not be confused with the Type Ia variety (neither by readers nor by the astronomers themselves!). Fortunately, the different supernova types can be distinguished spectrally. Supernova observations nowadays reach well beyond the megaparsec scales to the general-relativistic depths of the Universe, where we have to be a bit more specific about what we mean by distance itself.

Finally, it is worth mentioning that when we observe these further and further distances, we are also looking further and further back in time. This is even hard-wired into some of the distance units: one 'light year' is the *distance* that light travelled in a year, but that also means that the event took place *one year ago*. It is always the case that measurements and observations are made of past events, but perhaps in no other area of study is that felt more keenly than in astrophysics. We are almost directly watching events from hundreds of millions of years ago.

8.2 The Cosmological Principle

The *cosmological principle* is a formal version of the statement that there is nothing special about our location in the Universe. Well, one might point out that our location *is* special in many ways: we live on a planet with an atmosphere, and one that has continents and oceans, etc.; on our local scale, these all seem fairly unique. Zooming out a bit to a larger scale, we observe that we live in a galaxy some 20 kpc across. Looking further out, to $\sim$1 Mpc, we see another comparably large galaxy (Andromeda) and several small galaxies. Still further out, over tens of megaparsecs, there are some large clusters of galaxies (the Virgo and Coma clusters, APOD 110422 and APOD 150301). Over scales larger than $\sim$100 Mpc however, the Universe starts to to look pretty uniform, regardless of which direction our telescopes point, even though there are $\sim 10^{11}$ observable galaxies; these kinds of distances are what are generally referred to as the *cosmological scale*. On this basis, cosmology makes use of the following assumptions as an underlying principle:

I On large-enough scales, the Universe is *spatially homogeneous*, and in particular, it has uniform density. Properties may, however, be time-dependent.

II On large-enough scales, the Universe is *isotropic*, meaning that there are no special directions.

When this cosmological principle is combined with the additional assumption that general relativity is valid over cosmological scales, some interesting results follow. It is worth noting that, in any area of science, fundamental principles are always tested again and again by judging their consistency with both theoretical consequences and any observed phenomena; when mismatches occur, one must consider revising the principles or go back to a new start altogether. This is especially true in cosmology, where new observations have wrought paradigm shifts quite recently.

The first consequence of the cosmological principle is that spacetime on cosmological large scales is only allowed to curve in rather specific ways. In particular, curvature must be constant everywhere in accordance with the above principles of spatial homogeneity and isotropy; that is, properties of shapes and parallel lines cannot change simply by moving location. Such a spacetime is then describable by a metric of the general form (in Planckian units)

$$ds^2 = -dt^2 + a^2(t)\left[dr^2 + S^2(r)\left(d\theta^2 + \sin^2\theta\, d\phi^2\right)\right], \tag{8.2}$$

which is known as the Robertson–Walker metric (after H. P. Robertson and A. G. Walker, who studied its properties in the 1930s). This metric is in a polar form with coordinates (r, θ, ϕ), and since there is no preferred location, it is usual to place the observer (meaning ourselves) at the origin; the time coordinate t is known as *cosmic time.*

This metric in (8.2) resembles the Minkowski metric (3.8), but has two additional functions, $a(t)$ and $S(r)$. The factor $a(t)$ is known as the *scale factor*,[1] as it describes how distances stretch or shrink with time, while also accommodating the spatial homogeneity of the cosmological principle (note that it is a function of time only, and it multiplies all of the 'spatial' terms equally). It is common to define the current value of a as unity, writing $a(0) = 1$, and then one can compare the scaling factors at other times $a(t)$ directly.

The function $S(r)$ is perhaps an even more mind-bending factor: it changes the surface area of a sphere from $4\pi r^2$ to $4\pi S^2(r)$, while preserving isotropy (note that it is a function of length only, multiplies all of the 'angular' or 'directional' terms equally). In other words $S(r)$ expresses space curvature. For flat space, $S(r) = r$, so that it looks like a standard Cartesian-to-polar coordinate factor. Recent observations indicate that the Universe has $S(r) \approx r$, so we will keep this in mind for now, and postpone a discussion of curvature till Section 8.5.

Though space in the Universe may be (approximately) flat, spacetime itself is certainly curved. As noted earlier, this curvature is expressed in cosmology as the time-dependent scale factor $a(t)$. Spacetime is 'told' how to curve by the local energy density ρ. By the cosmological principle ρ must be independent of spatial coordinate, but can depend on time, and hence $\rho = \rho(t)$, consistent with the definition of its geometric 'partner' $a(t)$. Einstein's field equations for the metric reduce to a single equation relating the energy density and the scale factor (provided here in Planckian units):

$$\dot{a}^2 + k = \frac{8\pi}{3}\rho a^2\,, \tag{8.3}$$

known as the Friedmann equation. The dot over a denotes a derivative with respect to t, while k is an integration constant, which needs to be determined from observations. (The derivation of the Robertson–Walker metric and the Friedmann equation is given in most general-relativity texts. The abbreviation 'FLRW' is often used for the whole system, referring to Friedmann, Lemaître, Robertson, and Walker.)

It is important to point out that the scale factor itself does *not* imply that the Solar System or even individual galaxies will stretch of shrink with time. Recall that the cosmological principle applies only to regions large enough that underlying spatial density variations balance out to the cosmological average. Hence, the metric implied by the cosmological principle is not valid to be applied to the scale of a galaxy, or indeed on any scale where we can see extended structures.

Nevertheless, the scale factor is observable. That is thanks to a convenient consequence of general relativity: if the energy density in light is negligible compared to the

[1]This $a(t)$ is unrelated to the semi-major axis a of a Keplerian orbit.

overall density of the Universe, photons travelling through the Universe get their wavelength expanded with the Universe. In other words, for light emitted at wavelength λ and time t, the observed wavelength $\lambda_{\rm obs}$ at time t_0 is given by

$$\frac{\lambda_{\rm obs}}{\lambda} = \frac{a(t_0)}{a(t)} \,. \tag{8.4}$$

This relation is useful in probing the history of $a(t)$: if the particular atomic process giving rise to a spectral line λ can be inferred (e.g., through its relation to other spectral features), then by observing its present value $\lambda_{\rm obs}$, we can calculate the $a(t)$ from when the photon was emitted.

From the observations by Hubble and others *circa* 1930, a trend became noticeable: for any roughly standard candles, such as roughly similar galaxies, the fainter ones tended to have spectral lines shifted to longer wavelengths. That is to say, fainter-appearing standard candles were more redshifted. By the relation in (8.4), we can see that this leads to the conclusion that $a(t)$ has been increasing with time. Having an ever-enlargening scale factor implies that we live in an expanding Universe.

8.3 The Concordance Cosmology

In the Friedmann equation (8.3), the only two variables are the energy density ρ and the scale factor a. From this, we can see that understanding the $\rho(t)$ is key to understanding the expansion of the Universe. To proceed to do this, let us introduce the following quantity

$$H(a) \equiv \frac{\dot{a}(t)}{a(t)} \,, \tag{8.5}$$

which describes the relative rate of expansion at any epoch t; it has units of inverse time. Let us also make a simplifying assumption for now: in the metric (8.2) and the Friedmann equation (8.3), let us put that

$$S(r) = r, \quad k = 0 \,. \tag{8.6}$$

Current observations do indicate that these approximations are very good. The Robertson–Walker metric is then

$$ds^2 = dt^2 - a^2(t)\left(dr^2 + r^2 d\theta^2 + r^2 \sin^2\theta \, d\phi^2\right) , \tag{8.7}$$

and the Friedmann equation becomes

$$H^2(a) = \frac{8\pi}{3}\rho(a) \,. \tag{8.8}$$

Interestingly, looking at the units of variables here, we have yet again a relation where $1/\sqrt{\rho}$ is proportional to time. The expansion rate at the present epoch of the Universe, or $H(a=1)$, is known as the *Hubble constant* (or Hubble parameter) H_0. Its reciprocal ${H_0}^{-1}$ is known as the Hubble time and is the time scale parameter of the Universe.

Traditionally, the Hubble constant has been expressed as a speed divided by distance, whereas the Hubble time is simply a time. Measurements by different methods give

$$H_0{}^{-1} = 14 \pm 0.5\,\mathrm{Gyr}\,. \tag{8.9}$$

Referring back to the Friedmann equation (8.3), we can introduce that

$$\rho_0 = \frac{3}{8\pi} H_0{}^2\,. \tag{8.10}$$

This quantity is known as the cosmological critical density. Observations indicate that the average density of the Universe is very close to this value. In Planckian units

$$\rho_0 \simeq 1.8 \times 10^{-123}\,. \tag{8.11}$$

Of all physical quantities in Planckian units, the cosmological critical density is the most mind-bogglingly extreme, and is the origin of what is referred to as the '120 orders of magnitude problem' of popular science.

Let us look more closely at this energy density and what it is comprised of. One can express the cosmological critical density in other ways, such as baryon masses per cubic metre

$$\rho_0 \simeq 5.5\, m_\mathrm{b}\ \mathrm{m}^{-3}\,. \tag{8.12}$$

The mean density of stars and gas in the Universe is, however, only about 5% of this value. That is to say, at present baryonic matter (sometimes also called 'ordinary matter') accounts for only a small fraction of the total mass-energy of the Universe. That fraction of the cosmological energy density due to baryonic matter is conventionally written in equations as Ω_b (pronounced 'Omega-baryon', and it is unrelated to particles found in accelerator experiments whose names also use the same Greek letter). Present measures put it at $\Omega_b \simeq 0.045$.

What of the remaining 95% or so of the Universe's energy density? There are contributions from light and relativistic particles, which we will describe later. There is more non-relativistic matter in the Universe, though: non-ordinary, or *extra*ordinary matter, usually called *dark matter.* The name suggests something that absorbs light, but in fact dark matter appears to be transparent and also pressureless. The only reason (so far) we know that dark matter exists is that it interacts gravitationally and we see its effects. Recall from Chapter 1 the principle that orbital-speed squared times orbit size is proportional to mass. We saw this in the one- and two-body orbital velocity scale (1.8) and later in the N-body virial theorem (1.59). The same principle enables numerical models to infer mass from observations of kinematics, even when orbits are complex and complete velocity information is not available. Observations of the kinematics of gas in galaxies, pioneered in 1970s by Vera Rubin and Kent Ford, and by Albert Bosma and colleagues, leads to mass inferences much higher than the mass of stars and gas themselves. The difference is ascribed to dark matter. While the microscopic properties of dark matter are still conjectural, it appears to hang around galaxies and clusters of galaxies, rather than being spread more or less uniformly through space. Hence dark matter does not appear to be moving with relativistic velocities, and is dynamically pretty classical.

Returning to the question of energy density and cosmological scale factors, one can ask what happens to the relative contribution of constituents over time. The distances in the Universe are scaled by a (as defined in the Robertson–Walker metric), and therefore the matter per unit volume will be rescaled by a^{-3}. In other words the density of matter ρ_m (including both ordinary and dark) relative to the present energy density ρ_0 goes as

$$\rho_m(a) = \frac{\Omega_m}{a^3}\rho_0\,, \tag{8.13}$$

where Ω_m is the energy fraction of ρ_0 due to all matter (defined analogously to Ω_b, above).

For photons and other relativistic particles like neutrinos, the number density also goes as $\propto a^{-3}$, but there is a further effect to consider. This is that the wavelength itself changes $\propto a$ as the Universe expands, which happens according to the relation described in (8.4). As a result, the momentum per particle changes as $\propto a^{-1}$ (cf. Eq. (4.37)). For relativistic particles, energy varies like momentum; hence, the *relativistic* energy density of ρ_r picks up an extra factor of a relative to nonrelativistic mass, and

$$\rho_r(a) = \frac{\Omega_r}{a^4}\rho_0\,. \tag{8.14}$$

Here we have introduced a new dimensionless parameter Ω_r, the energy fraction of ρ_0 due to relativistic substances. (We reserve Ω_γ to refer specifically to radiation.)

Another contributor to the density (in fact, the largest one at the present epoch) is the *cosmological constant* Λ. Unlike the other contributing terms, this one is independent of a, and the effective density due to Λ has the form

$$\rho_\Lambda = \Omega_\Lambda\,\rho_0\,. \tag{8.15}$$

Although the presence of such a thing totally violates classical conservation laws, it manages to be consistent with general relativity by having negative pressure. (In fact, speculation of the existence of something like Λ goes back to Einstein.) Λ is called *dark energy,* especially in popular science. But the descriptor 'dark energy' completely understates the weirdness of Λ, and it also runs the risk of confusion with the (much less exotic) dark matter. 'Dark tension' has been suggested as a more descriptive term, because of Λ's defining characteristic of acting as negative pressure; however, even dark tension is not satisfactory, since the tension causes more expansion. It seems Λ is so weird, it defies description.

There could be even more ghostly things that behave like densities contributing to the overall ρ. For example, a certain kind of topological defect in spacetime has been hypothesized, known as a 'texture', which would give an effective density $\propto 1/a$. A possible 'phantom' density $\propto a$ has also been mentioned occasionally in the research literature. Let us continue, however, without further exotic density components, as there is no compelling evidence for them as yet.

As the fractional densities have been defined, they add up to unity

$$\Omega_m + \Omega_r + \Omega_\Lambda = 1\,, \tag{8.16}$$

so that the Friedmann equation (8.8) becomes

$$H^2(a) = H_0{}^2 \left(\frac{\Omega_m}{a^3} + \frac{\Omega_r}{a^4} + \Omega_\Lambda \right). \tag{8.17}$$

Current observations, some of which we will explain later, give the following values of the cosmological parameters:

$$\begin{aligned} \Omega_m &\simeq 0.3 \quad \text{(which includes } \Omega_b \simeq 0.045\text{)}, \\ \Omega_r &\simeq 8 \times 10^{-5} \quad \text{(which includes } \Omega_\gamma \simeq 5 \times 10^{-5}\text{)}, \\ \Omega_\Lambda &= 1 - \Omega_m - \Omega_r, \\ H_0{}^{-1} &\simeq 14\,\text{Gyr}. \end{aligned} \tag{8.18}$$

These make up what is often called the *concordance model* of cosmology.

The Friedmann equation (8.17) with its multiple components (8.18) requires a computer to solve. Sometimes it is convenient to work with a simpler case, the so-called Einstein–de Sitter Universe, which has $\Omega_m = 1$ and the other Ω all zero, leading to

$$H(a) = H_0\, a^{-3/2}. \tag{8.19}$$

Historically speaking, the Einstein–de Sitter cosmology was a highly favoured model, in large part because of its elegance and simplicity. However, since around just the start of the 21st century, observational advances have pushed a paradigm shift to the concordance cosmology.

Exercise 8.1 While H_0 and the Ω parameters are constants over the history of the Universe, they are in a sense anchored at our epoch. Imagine astrophysicists at $a = 0.5$ of the concordance cosmology. They prefer to define a as unity in their own epoch. What numbers would they obtain for Ω_m and Ω_Λ? *Hint:* Work out the densities at their epoch first.

Exercise 8.2 Show that dark energy amounts to energy generation at the rate of $9/(8\pi) \times H_0{}^3\,\Omega_\Lambda$ per unit volume. For a volume the size of the Sun (the Sun has radius $\simeq 2$ light-sec), show that the energy-generation rate would be about 3 watts.

8.4 Distances and Lookback Times

When observing distant objects, we must keep in mind that while the photons are travelling toward us the Universe is itself expanding. Our normal ideas of observation, time, and distance must be adapted accordingly. Suppose a photon was emitted at cosmic time t in the past and is then detected by us now (at cosmic time zero). This interval, which is the amount of time that a photon took to reach us, is known as the *lookback time*. The lookback time can naturally be thought of as a distance, but is different from the current distance or radial coordiate r of the emitting object, because the Universe has been expanding while the light has been travelling. Both t and r can be related to the scale factor $a(t)$ when the light was emitted, which is useful because

$a(t)$ can be measured directly via Eq. (8.4). The relations can be stated as two little differential equations:

$$\begin{aligned} \frac{dt}{da} &= \frac{1}{aH(a)}\,, \\ \frac{dr}{da} &= \frac{1}{a^2\,H(a)}\,. \end{aligned} \tag{8.20}$$

The equation for t is just a rewritten form of the definition (8.5) of the Hubble expansion rate. To derive the equation for r, recall that photons making the journey to us move along geodesics, i.e., spacetime paths with $ds = 0$. From the metric (8.2) we have $dt = a(t)\,dr$. Combining this with the equation for $t(a)$ gives the equation for $r(a)$. Initial conditions for the observer, which are really 'final conditions' for the photon, are $t = 0$, $a = 1$, $r = 0$.

Eqs. (8.20) can also be written as integrals. For the Einstein–de Sitter case, the integrals are tractable analytically (see Exercise 8.4 at the end of this section). The concordance cosmology, however, requires numerical solution. We have already met several examples of differential equations requiring numerical solution in previous chapters, and Eqs. (8.20) are not insurmountable, either. Figure 8.1 shows $r(a)$ and $t(a)$.

Lookback time would be an observable if there were clock-like processes running steadily since the early Universe, but in practice lookback time is inferred as $t(a)$ via a cosmological model. Thus, as we can read directly from the lower panel of Fig. 8.1, the age of the Universe today is nearly ${H_0}^{-1}$ in the concordance model but less in the Einstein de Sitter model. If we still want to visualize $a(t)$, we could just mentally rotate the lower panel, which describes $t(a)$, counterclockwise by 90°. We could then say that the early Universe expanded very fast, but then decelerated. In the Einstein de Sitter model, the deceleration gradually weakens but continues to the present. In the concordance model, the deceleration stops by $t \simeq 0.5{H_0}^{-1}$ and changes into a slight (imperceptible in the figure) acceleration.

Likewise, distance $r(a)$ is a quantity inferred from a by way of a cosmological model. As Fig. 8.1 shows, for small a, $r(a)$ can be much larger than the age $t(0)$ of the Universe. Thus, r is a 'notional distance' rather than an observable quantity, and one is free to define other notional distances. Another such case is the so-called 'luminosity distance' D_{lum}, defined such that the observed brightness of standard candles goes as D_{lum}^{-2}. Why is D_{lum} not just r? Because of cosmological redshift: photons coming from a source at a are redshifted by a factor of $1/a$, meaning that (a) they arrive a times as frequently and (b) they are each a as energetic. The energy reception rate (i.e., apparent luminosity) is thus a^{-2} of what it would have been from the same r, had there been no cosmological redshift. Accordingly,

$$D_{\text{lum}} = \frac{r}{a}\,. \tag{8.21}$$

Another kind of distance comes from considering the apparent size of objects with the same intrinsic size according to 'standard rods'. Now, although light gets redshifted as it travels through the expanding Universe, light *paths* (i.e., the geodesics) are not affected by the expansion. Thus, the apparent angular sizes are the same as they would have been if there had been no expansion between emission and observation. (This

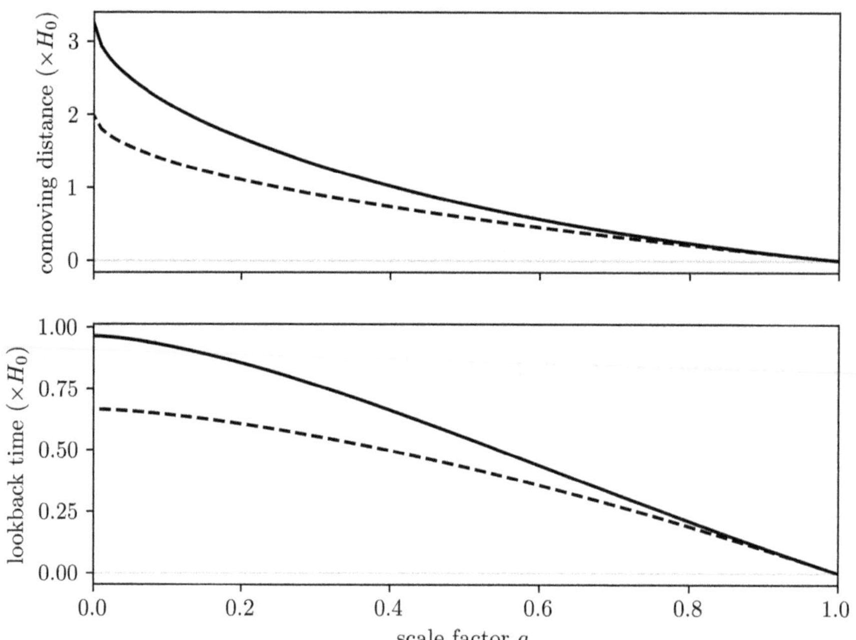

Fig. 8.1 Distance and lookback times, in units of ${H_0}^{-1}$, in the concordance cosmology (solid curves) and the Einstein de Sitter cosmology (dashed curves). The present epoch is at the right ($a = 1$), and the Big Bang is to the left.

would be disregarding separate bumps along the way, that is, gravitational lensing by localized gravitational fields.) When the light was emitted, distances were a times what they are now. Accordingly, the angular-diameter distance is defined as

$$D_{\mathrm{ang}} = ra\,. \tag{8.22}$$

Thus far we have expressed distances and lookback times in terms of a. In place of a, one can also use the amount of observed redshift as the independent variable, which is known as the cosmological redshift and written using the unitless variable z. Its definition is based on Eq. (8.4) with $a = 1$ where the observation occurs, as

$$1 + z \equiv \frac{1}{a} = \frac{\lambda_{\mathrm{obs}}}{\lambda}\,. \tag{8.23}$$

This is analogous to the kinematical redshift seen with Eq. (1.61) in Section 1.11. But whereas the kinematic redshift formula (1.61) is classical and applies only to velocities $\ll c$, the cosmological redshift formula (8.23) can be used for any redshift, even $z \gg 1$.

Lookback times and distances are often written as integrals over z. To do so, we write the scale factor as a function of redshift, with $a = a(z) = 1/(1 + z)$ and

$da = -a^2\,dz$ in Eqs. (8.20), which can then be rewritten as integrals. The luminosity distance (8.21) can then be expressed as

$$D_{\text{lum}} = (1+z)\int_0^z \frac{dz'}{H(a(z'))}\,. \tag{8.24}$$

A measurement of apparent brightness for a standard candle gives that object's D_{lum}. A plot of this redshift against luminosity distance is known as a *Hubble diagram.*

For $z \ll 1$ (which refers to our relative neighbourhood, on a cosmic scale), the redshift–distance relation (8.24) reduces to the linear approximation

$$H_0\,D_{\text{lum}} \approx cz\,. \tag{8.25}$$

(We have put back the formerly implicit factor of c here.) This linear relation is known as Hubble's law, and is a description of the popular view of the expanding Universe: the right-hand side is interpreted kinematically as a recession velocity, and we have galaxies flying away from us with a speed proportional to their distance. Cosmological distances have traditionally been given in megaparsecs, and the Hubble constant is often given as

$$H_0 = 100\,h\ \text{km}\,\text{s}^{-1}\,\text{Mpc}^{-1}\,, \tag{8.26}$$

with the dimensionless factor $h \simeq 0.7$.

The linear form of Hubble's law in (8.25) is no longer used in cosmology. Galaxies are now identifiable far beyond $z = 1$ (for examples of galaxies up to $z \simeq 8$, see APOD 130827), which goes far beyond the realm where the linearized approximation is valid. Therefore, it is essential to use the more general form (8.24) of the Hubble diagram relation. The units for the Hubble constant in (8.26) are, however, still widely used.

Exercise 8.3 Find the value and uncertainty in h implied by the Hubble-time value and uncertainty given in Eq. (8.9). (The precise value of h is much debated.)

Exercise 8.4 For Einstein–de Sitter, find expressions for lookback time and the luminosity and angular-diameter distances as functions of a. Show further that (a) the time since $a = 0$ is $\frac{2}{3}\,{H_0}^{-1}$, and (b) D_{ang} is not monotonic, but has a maximum at $a = 4/9$.

Exercise 8.5 Using a numerical solver for (8.20), compute the luminosity distance in the concordance cosmology. Try out different values for the Ω constants, while keeping the sum unity (Eq. 8.16). You should see that introducing Ω_Λ makes the luminosity distance increase faster with redshift. This effect was first measured in the late 1990s from Type Ia supernovae by two independent teams, one led by B. Schmidt and A. Riess, and the other by S. Perlmutter.

8.5 Curvature and its Consequences

In Section 8.2 we briefly introduced the curvature parameter k and then promptly set it to zero. We now consider the consequences of non-zero k.

It is convenient to move k to the right-hand side of the Friedmann equation (8.3), and treat it formally like another contribution to the density. Accordingly, let us introduce Ω_k, which again is a fractional contribution to energy density, and such that

$$k = -{H_0}^2\,\Omega_k\,, \quad \Omega_m + \Omega_r + \Omega_\Lambda + \Omega_k = 1\,. \tag{8.27}$$

We remark that, unlike the other fractional Ω terms, Ω_k can be positive or negative. Combining these relations with Eqs. (8.3) and (8.5), analogous the the derivation of Eq. (8.17), the Friedmann equation now becomes

$$H^2(a) = {H_0}^2 \left(\frac{\Omega_m}{a^3} + \frac{\Omega_r}{a^4} + \Omega_\Lambda + \frac{\Omega_k}{a^2} \right)\,. \tag{8.28}$$

As a result, the expressions in Eqs. (8.20) and (8.22) for lookback time and distances now become dependent on k, as well.

The really striking effect of k, however, is on the metric in Eq. (8.2), through the $S(r)$ factor. Depending on the sign of k, $S(r)$ takes different functional forms, as follows:

$$S(r) = \begin{cases} \dfrac{\sin(\sqrt{k}r)}{\sqrt{k}} & k > 0 \quad \text{‘spherical’}, \\ r & k = 0 \quad \text{‘Euclidean’}, \\ \dfrac{\sinh(\sqrt{-k}r)}{\sqrt{-k}} & k < 0 \quad \text{‘hyperbolic’}. \end{cases} \tag{8.29}$$

These expressions encode space curvature. For $k > 0$, space is like the three-dimensional surface of a sphere in four dimensions. For a lower-dimensional analogy, think of the world map on the United Nations emblem. The r coordinate corresponds to the radius of the circles on the map. The actual circles on the Earth do not grow as fast as on the map, and south of the equator the circles on the Earth get smaller even as the circles on the map get bigger. This effect is accounted for by the $S(r)$ factor. The opposite case, or $k < 0$, corresponds to hyperbolic geometry. There is no familiar analogy for this case, but it has been explored in art, notably in the *Circle Limit* prints by M. C. Escher.

Before the concordance cosmology became widely accepted (around 2000), a cosmology with just matter and curvature, having $\Omega_\Lambda = \Omega_r = 0$ and $\Omega_m = \Omega$, $\Omega_k = 1-\Omega$, was much studied. Putting these parameters into the Friedmann equation (8.28) gives

$$\dot{a}^2 = {H_0}^2 \left(\frac{\Omega}{a} - (\Omega - 1) \right)\,. \tag{8.30}$$

This equation has the same form as the energy in Eq. (1.28) for Keplerian orbits (with a here as r there, and the remaining quantities merely constants). Because of this Newtonian analogy, the matter-only cosmology remains of some interest.

Exercise 8.6 Solve for $a(t)$ in a curved matter-dominated cosmology given in Eq. (8.30). The result has two different forms for $\Omega < 1$ and $\Omega > 1$.

Exercise 8.7 In the text we have taken r as an independent variable and $S = S(r)$. We could also take S as an independent variable and $r = r(S)$. If this is done, show that the metric in Eq. (8.2) becomes

$$ds^2 = -dt^2 + a^2(t)\left[\frac{dS^2}{1 - k\,S^2} + S^2\left(d\theta^2 + \sin^2\theta\,d\phi^2\right)\right].$$

Exercise 8.8 In older literature one often finds the approximation

$$a(t) = 1 + H_0\,t - \tfrac{1}{2}q_0 H_0{}^2\,t^2,$$

where t is measured with respect to the present epoch. The constant q_0 is known as the deceleration parameter, and appropriately it is negative in the concordance model. Show that

$$H(a(z)) = H_0\left(1 + (1 + q_0)z + O(z^2)\right).$$

This approximation applies only for small z, and hence $H(a(z))$ is only slightly nonlinear. In this regime Ω_r is negligible. In this regime, show that

$$q_0 = \tfrac{1}{2}\Omega_m - \Omega_\Lambda,$$

where Ω_k has been replaced by $1 - \Omega_m - \Omega_\Lambda$.

Exercise 8.9 In terms of q_0 as defined in Exercise 8.8, show that the luminosity distance in Eq. (8.24) goes nonlinear as

$$D_{\text{lum}} = H_0^{-1}\left(z + \tfrac{1}{2}(1 - q_0)z^2 + O(z^3)\right).$$

8.6 Standard Sirens

Since the Big Bang itself, the most cataclysmic events in the Universe have been the mergers of black holes. These events outshine (briefly) the entire rest of the observable Universe. The trouble with them has been that they shine in gravitational radiation, and nobody could see them—until 2015. APOD `160211` summarises the long-sought-after first detection of gravitational waves from a merging black hole binary. APOD `160615`, which describes the second confirmed gravitational wave source, exhibits an interesting feature of the waves from mergers: the frequency and amplitude of the waves both increase just at the end of the event. The frequency increase is often described, invoking sonic metaphors, as a siren or chirp.

Gravitational waves are analogous to density waves in a medium, with the 'space' part of the metric playing the role of the medium. One viewable effect they have is to change the phase of light waves they encounter, which makes them detectable using laser interferometers. To formulate these statements precisely would need a lot more machinery from general relativity than we can cover here, and we will not attempt to do so. Nevertheless, we would like to understand the broad features of the waves, such as why the characteristic signal has a 'siren' or chirping property, as noted above

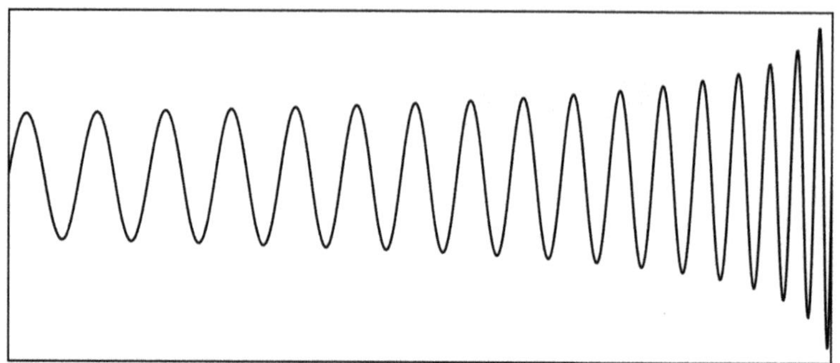

Fig. 8.2 The waveform of a standard siren gravity wave, from Eq. (8.42). The horizontal axis is time and the vertical axis is the observable (called the *strain*), but the units are arbitrary.

and also illustrated in Fig. 8.2, and what this tells us. So we will take a semi-classical approach, adding relativistic waves to a classical system in ad hoc fashion, and then follow up the consequences.

We start by considering two *classical* bodies in a circular orbit. Let their masses be m_1 and m_2 and the orbital radius be $a_{\rm orb}$. In reality they would be relativistic and have perturbed Schwarzschild metrics around them (perturbed because they interact with each other), but we will disregard relativity for now, and approximate the system as classical. The orbital energy E and orbital period P of this system are given (in Planckian units) by

$$E = -\frac{m_1 m_2}{2a_{\rm orb}}, \qquad P^2 = (2\pi)^2 \frac{a_{\rm orb}^3}{m_1 + m_2}. \tag{8.31}$$

The orbital period is taken from Eq. (1.8), remembering the result from Section 1.7 that we have to use the total mass. The orbital energy comes Eq. (1.29); that expression is actually for energy per unit mass, but the result of Exercise 1.10 tells us that the relevant mass to use is the reduced mass (1.42). We next introduce two curious-looking quantities

$$M^5 \equiv \frac{(m_1 m_2)^3}{m_1 + m_2}, \qquad q^5 \equiv \frac{m_1 m_2}{(m_1 + m_2)^2}\, a_{\rm orb}^5 . \tag{8.32}$$

Here M is a sort of weighted mean of the masses, and q is the orbital radius scaled by a factor that also depends on the mass ratio. The primary advantage of M and q is that we can write the orbital energy and period rather simply as

$$E = -\frac{M^2}{2q}, \qquad P^2 = (2\pi)^2 \frac{q^3}{M}. \tag{8.33}$$

We now posit a *non-classical* process, namely that the system emits waves. We also assert that their amplitude h is proportional to the system's gravitational energy E

and that it fades with distance D according to an inverse-linear relationship (discussed further in the next paragraph), namely that:

$$h(D) \propto \frac{E}{D}. \tag{8.34}$$

The amplitude h is dimensionless and corresponds to a fractional stretching or shrinking of space (a highly non-classical process!). We approximate that the waves are emitted isotropically, which is not the case for real gravitational waves, but it is good enough for our present purposes. The emission frequency is twice the orbital frequency, that is, the period of the gravitational wave is $P/2$. To see why the period should be halved, consider an equal mass binary: after half an orbit, the masses will have swapped position, but the mass distribution will be like before. For unequal masses, the argument is more subtle (see below) but the result is actually the same.

Why would the amplitude be modelled as proportional to the orbital energy? One view is to take h as being analogous to a potential in ordinary physics (e.g., $V \propto M/r$ for Newtonian gravity or $V \propto e/r$ for Coulombian charges) but with the system's E in the numerator. Alternatively, we may borrow from general relativistic treatments, where the amplitude is shown to be proportional to the second derivative of the moment of inertia: $h \propto \ddot{I}/D$. Back in Section 1.8, we derived the moment of inertia of a binary as $I = M_{\rm red}\, a_{\rm orb}^2$ or $I = m_1 m_2/(m_1 + m_2) \times a_{\rm orb}^2$. In terms of the quantities in Eq. (8.32), here we would then have $I = Mq^2$. Importantly, there is an accompanying angular factor oscillating twice per orbit (cf. Exercise 1.11). Upon including the angular dependence, the second derivative becomes $\ddot{I} = (\pi/P)^2 I$. Substituting from the orbital relations (8.33) we have $\ddot{I} \propto E$. Hence the justification for $h(D) \propto E/D$.

Continuing semi-classically, we assume that the h wave carries energy like other waves, with energy density proportional to (amplitude $\times$ frequency)^2. Now consider a sphere of radius D around the binary. The rate of energy transported through this surface by the wave equals the energy density of the wave, times its speed, times the area of the surface. The wave travels at light speed, thus unity in Planckian units. From the two remaining factors, we see that the rate of energy $\dot{E}$ coming out of this sphere will be $\propto D^2(h/P)^2$. Substituting from Eq. (8.34) into $E \propto D\, h(D)$, we have

$$\dot{E} \propto \frac{E^2}{P^2} \tag{8.35}$$

for the rate of energy emission, or power.

Having inserted a semi-classical energy-loss process into the two-body problem, we now consider the classical consequences. The energy source for the gravitational wave can only be the orbital energy of the two bodies. The loss of orbital energy $\dot{E}$ will make the orbital radius smaller. Consequently, the period will shorten at some rate $\dot{P}$. Since E and P are both proportional to various powers of q, we can write

$$\frac{\dot{P}}{P} \propto \frac{\dot{E}}{E}. \tag{8.36}$$

We now have a total of eight quantities, and five relations connecting them: the two-body parameters (M and q) and the distance (D) are basic; using them, we define the further classical quantities E and P in Eq. (8.33) and the non-classical quantities h, $\dot{E}$, and $\dot{P}$ in Eqs. (8.34) and (8.36). In gravitational-wave observations of a binary such as in APOD 160211, observers use an interferometer to measure the oscillation's amplitude h and period $P/2$. Furthermore, they track how the frequency of the oscillations increases as the system gradually approaches merger, thus measuring $\dot{P}$. From these three observables $(h, P, \dot{P})$, we can now solve for the remaining five quantities $(\dot{E}, E, D, M, q)$.

Squaring Eq. (8.36) and then substituting Eq. (8.35) for one of the $\dot{E}$ factors gives

$$\dot{E} \propto \dot{P}^2 \,. \tag{8.37}$$

Substituting the last expression for $\dot{E}$ in Eq. (8.36) gives

$$E \propto P\dot{P} \,. \tag{8.38}$$

Comparing this expression with the classical relations in Eq. (8.33) gives

$$M^5 \propto P^5 \dot{P}^3 \,. \tag{8.39}$$

The weighted-mean mass M is called the 'chirp mass' because it can be inferred from the frequency-shift of the waves. For the black hole merger in APOD 160211, $P \simeq 0.06\,\mathrm{s}$ as the signal first became significant, and quickly shortened at $\dot{P} \simeq 0.2$. (Note that we are referring here to the orbital period, which is twice the period of the gravitational-wave signal.) If we plug these numbers into the mass formula in Eq. (8.39), that greatly overestimates the mass (a detailed relativistic model gives $M \simeq 30M_\odot$, as widely reported). Clearly, the proportionality constant in Eq. (8.39) is far from 1. The classical relation in Eq. (8.33) suggests that we could do better by using $P/2\pi$ and $\dot{P}/2\pi$, and indeed this improves the approximation. Relativistic perturbative theory supplies a further factor of $(5/96)^3$ to the formula in Eq. (8.39).

Inserting the expression for E from Eq. (8.38) in the amplitude formula in Eq. (8.34) gives

$$hD \propto P\dot{P} \,. \tag{8.40}$$

Apart from a numerical factor, which we will not go into here, this relation gives the distance in terms of the observables h, P, and $\dot{P}$. Finally, substituting the classical relations in Eq. (8.33) into the radiated power gives

$$\frac{M^5}{q^5} \propto \dot{E} \,, \tag{8.41}$$

which determines q.

We can now determine the time dependence of the observables. To do so, we treat Eq. (8.39) as a little differential equation for P. Proceeding to solve for P as a function of time, and then putting the result into Eq. (8.40) gives

$$P \propto (-t)^{3/8} \,, \qquad h \propto (-t)^{-1/4} \,, \qquad \text{signal} \propto h \times \cos(4\pi t/P) \,. \tag{8.42}$$

Figure 8.2 illustrates. At $t = 0$ formally the period becomes zero, and h becomes infinite, but this is in the future; hence, t is considered negative. The $P \to 0, h \to \infty$

event does not really happen, because the system changes from orbiting to merging before $t = 0$. The binary model breaks down as the masses merge, but the merger can still be followed using numerical general relativity, enabling still more information, including the original masses m_1 and m_2, to to be reconstructed.

Binary systems emitting gravitational waves with period $\frac{1}{2}P$, while P gradually decreases as the orbital energy is radiated way, must be a rather common phenomenon in the Universe. Because of their simple outward properties, these creatures are sometimes called 'standard sirens'. The masses need not be black holes, and indeed there are binary white dwarfs in the Milky Way with known M, P, and D whose gravitational waves are expected to be detected by future laser interferometers, and would be test cases (or verification binaries) for such detectors. And actually, the first gravitationally radiating system was discovered in 1974! It is a neutron-star binary named after its discoverers as the Hulse–Taylor pulsar. The h from it is too small to measure, but J. Taylor and collaborators measured the system's orbital energy by other methods and over two decades showed that its $\dot{P}$ agreed precisely with what is expected from gravitational radiation.

Exercise 8.10 Suppose we scale up a given black hole binary. That is, we increase M and q by the same factor, and keep D the same. How will that affect the other quantities?

Exercise 8.11 Show that the output power by a standard siren, as it approaches merger, will be of order unity (or c^5/G in non-Planckian units), no matter what M is. Then consider a Hubble volume, that is, a sphere of comoving radius $1/H_0$. If the entire baryonic mass in a Hubble volume radiates with the Eddington luminosity (a considerable overestimate) what is the power output?

Exercise 8.12 The distance in Eq. (8.40) will be the luminosity distance, assuming the observable $P, \dot{P}, h$ factors are used. That is, a factor of $1 + z$ corresponding to the emission epoch is somehow implied. Why is that?

Exercise 8.13 One can think of $E/\dot{E}$ as a time scale for a binary system to shrink towards merger. Estimate this time scale for the Hulse–Taylor binary, where P is about 8 hr and M is about a solar mass.

8.7 Redshift Drift

So far, we have tacitly assumed that all observations happen at one time. What if we observe over a period of time?

The measured redshift of an object will be

$$1 + z = \frac{a(t_{\rm obs})}{a(t)}, \tag{8.43}$$

where t is the emission time for a photon observed at $t_{\rm obs}$. Now consider the difference in redshift at a slightly later time:

$$\Delta z = \frac{a(t_{\rm obs} + \Delta t_{\rm obs})}{a(t + \Delta t)} - \frac{a(t_{\rm obs})}{a(t)}. \tag{8.44}$$

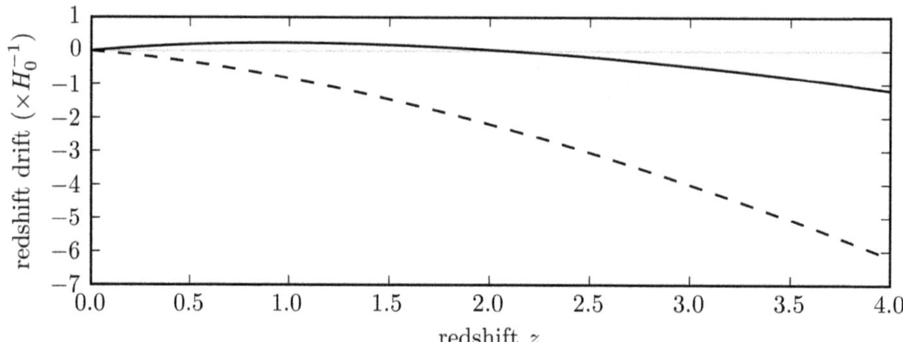

Fig. 8.3 Redshift drift $dz/dt_{\rm obs}$ in units of H_0, in the concordance cosmology (solid curve) and the Einstein de Sitter cosmology (dashed curve).

The redshift of photons is equivalent to a time dilation, and thus

$$\frac{\Delta t}{\Delta t_{\rm obs}} = \frac{a}{a_{\rm obs}}, \tag{8.45}$$

defining $a_{\rm obs} \equiv a(t_{\rm obs})$. Substituting into (8.44) and simplifying gives

$$\frac{dz}{dt_{\rm obs}} = \frac{\dot{a}_{\rm obs} - \dot{a}}{a} = \frac{H_0}{a} - H(a)\,. \tag{8.46}$$

This effect is known as *redshift drift.*

Figure 8.3 shows redshift drift for the concordance and the Einstein de Sitter cosmologies. It is expressed as a function of the (undrifted) redshift $z = 1/a - 1$. Redshift drift is a *very* small effect, whose scale is set by $H_0 \sim 10^{-10}\,{\rm yr}^{-1}$ and observing it is still a dream. Nonetheless, some observational programs are attempting to measure redshift drift over a period of several years, by accumulating data on a very large number of objects. If they succeed, they will unravel the history of the acceleration and deceleration of the Universe directly, rather than having to infer it indirectly.

9
The Cosmic Microwave Background

The theory of the expanding Universe implies that once upon a time, roughly 14 billion years ago, the scale factor of the Universe was zero. The commonly used term 'Big Bang' refers to this singular point. Standard physics cannot say anything about the Big Bang singularity—how can one cope with zero lengths and infinite densities? It is possible to enlarge standard physics—for example, with conjectural extra dimensions—and then it does become possible to describe what may have happened exactly at the Big Bang, and even before (if there was a before). Perhaps one of the beyond-standard-physics theories in the research literature is correct, or maybe several are close. We can only guess, at this stage.

Nevertheless, well-understood physics can be used to study processes that happened quite soon after the Big Bang. It was Gamow (fittingly, a past student of Friedmann) who realized this in the 1940s. Importantly, these processes are still directly observable even today.

The key quantity to consider is the radiation density. As we saw in the previous chapter, the ambient radiation in the Universe has a relatively tiny energy density, which makes it dynamically unimportant today. But it was not always so. Recall that the expansion of the Universe does not drive the addition or removal of photons, but it does tend to stretch photon wavelengths through the scale factor. In the past the ambient photons (not photons emitted recently) would have had shorter wavelengths and hence been more energetic. This would have given the Universe a higher ambient temperature than it has today, and if you go *far enough* back in time, then there would have been high-energy microphysical processes occurring everywhere in this extreme environment.

Gamow studied nuclear processes in the hot early Universe, especially those of helium. When the temperature was high enough, atomic nuclei would have been constantly forming and breaking up, and a dynamic equilibrium between hydrogen and helium nuclei would have existed. Then, as the Universe expanded and the ambient radiation cooled, the reactions would have stopped rather abruptly, leaving a quarter or so of baryons locked up in helium nuclei, which is what we find today. This topic of what amounts of various elements were formed during this period is known as Big Bang nucleosynthesis; the physics of this time determined what the contents of the early Universe were—it is a large and fascinating topic in its own right. Here, we will just note that Gamow, with his younger colleagues R. A. Alpher and R. C. Herman, developed an early form of it from 1948 onwards, and that one of their results was to estimate the current temperature of the Universe. Within their various models, their estimates varied within the range of $T = 5$ to $50\,\mathrm{K}$.

The Astronomer's Magic Envelope. Prasenjit Saha and Paul A. Taylor.

10.1093/oso/9780198816461.001.0001

Actually, a temperature measurement of about 2.3 K for interstellar gas had already been made in 1941 by A. McKellar. But, as it happens, nobody made the connection at the time. Then in 1964 A. A. Penzias and R. A. Wilson detected a mysterious microwave signal coming from all directions, ultimately showing that the sky is filled with a photon gas at $T = 2.7\,\mathrm{K}$, after which the pieces of the cosmic puzzle finally fell into place.

In the early years of the 21st century, detailed study of the cosmic microwave background (CMB) has been one of the most active areas of astrophysics. Let us try and get a feeling for why.

9.1 Radiation Density and Matter Density

The microwave background is a near-perfect example of a theoretical photon gas (recall Section 4.5), far closer, in fact, than any laboratory realisation on Earth. The temperature has been measured to be

$$T_{\rm cmb} = 2.725\ \mathrm{K}\,, \tag{9.1}$$

and more generally,

$$T = T_{\rm cmb}/a\,. \tag{9.2}$$

Recalling the number and energy densities of a photon gas in Eqs. (4.34) and (4.35) gas, we have

$$\begin{aligned} n_\gamma &= 0.2435\,(T_{\rm cmb}/a)^3\,, \\ \rho_\gamma &= \frac{\pi^2}{15}(T_{\rm cmb}/a)^4\,. \end{aligned} \tag{9.3}$$

The number density comes to 410 photons cm^{-3} at the present epoch of the Universe. The parameter Ω_γ from the previous chapter is just ρ_γ/ρ_0.

The ratio of baryons to photons can be inferred from the study of different nuclear processes. For example, the higher that ratio, the more ^{4}He would have been produced in the early Universe. The so-called primordial helium abundance can be measured from the absorption spectra of intergalactic gas, which has not been enriched with helium from stars. We will not go further into nucleosynthesis in this book, but simply take over the result

$$\eta \equiv \frac{n_{\rm b}}{n_\gamma} \simeq (5.5 \pm .5) \times 10^{-10}\,. \tag{9.4}$$

Converting to energy densities, we get

$$\frac{\rho_{\rm b}}{\rho_\gamma} = \frac{\eta\, m_{\rm b}\, n_\gamma}{\rho_\gamma} \simeq 800\,a\,. \tag{9.5}$$

In other words, radiation matched the energy density of baryons at $a \simeq 1/800$. Since the total matter is about six times the baryonic matter, matter–radiation equality was earlier at $a \simeq 1/5,000$.

Exercise 9.1 Since the Sun keeps us a lot warmer than 2.7 K, we might expect that stars dominate the ambient radiation. But can they really? How far from the Sun does its photon-number flux equal that of the CMB? How far away do the energy fluxes become equal? (The Sun has a radius of about 2 light-sec and a surface temperature around $2 \times 10^3\, T_{\rm cmb}$.)

9.2 Recombination

Hydrogen gas is pretty transparent throughout most of the electromagnetic spectrum. At longer wavelengths than ultraviolet, only some narrow lines in the spectrum are absorbed. However, in the ultraviolet range photons can dissociate molecules and ionize atoms, making hydrogen opaque, because free electrons can scatter photons of all wavelengths—recall the discussion of the Thomson cross-section σ_{T} and its expression in Eq. (7.10).

We can quantify opacity using

$$\tau = \int n_e \, \sigma_{\mathrm{T}} \, dt \,, \tag{9.6}$$

where n_e is the number density of free electrons and the integral performed is along a light path. The dimensionless number τ (unrelated to *other* quantities also called τ in this book) is known as the *optical depth.* Essentially, it provides a measure for how far photons would be expected to travel before being absorbed—and therefore providing a measure of how far into a substance we would be able to see. As a terrestrial example, it is analogous to the 'visibility' distance that is reported on foggy days.

Let us see what happens for the 'maximal' scenario that baryonic matter is entirely in the form of ionized gas. Taking the baryonic component of the matter density (8.13), we have $\rho_{\mathrm{b}}(a) = \Omega_b \rho_0 / a^3$. Using Eq. (8.10) to eliminate ρ_0, we have $\rho_{\mathrm{b}}(a) = 3\Omega_b {H_0}^2/(8\pi \, a^3)$. Assuming the gas is ionized hydrogen and has an equal number of protons and electrons, the free-electron density is

$$n_e = \frac{3}{8\pi} \frac{\Omega_b {H_0}^2}{m_{\mathrm{b}} a^3} \,. \tag{9.7}$$

Inserting this expression for n_e, together with formula (7.10) for σ_{T}, into Eq. (9.6), and further changing variables from t to a (cf. Eq. 8.5), gives the following expression for optical depth:

$$\tau = \frac{\alpha^2}{m_e^2 m_{\mathrm{b}}} \, \Omega_b H_0 \int \frac{H_0}{a^4 H(a)} \, da \,. \tag{9.8}$$

This expression provides an optical depth within our own Universe—from how far away (i.e., from how far back in time) can we expect to collect photons? The answer depends on the particular cosmological model, codified by what goes into $H(a)$. Here the pre-factor, for which we must be careful to convert H_0 to Planckian units, comes to $\simeq 0.002$, while the integral evaluates to a dimensionless factor, since a is just a number and the units of H cancel. For most cosmologies of interest, the integral itself requires a numerical solution. But in the simpler Einstein–de Sitter case in Eq. (8.19), we get

$$\tau \simeq 10^{-3} \left(a^{-3/2} - 1 \right) \,. \tag{9.9}$$

Thus, in an ionized Einstein–de Sitter Universe we could look back to $a \sim 10^{-2}$.

In reality the state of the intergalactic medium is actually a little more complicated. When the Universe was very hot, intense radiation bathed everything and nuclei were unable to hold onto electrons; the gas was maintained in a state of full ionization.

Then, as the Universe cooled by expansion, the balance tipped towards electrons being captured in bound states around nuclei, and the gas became atomic. That episode is known as recombination. Much later, when stars and galaxies formed, a 'reionization' occured in the intergalactic gas, whereby electrons became unbound from hydrogen nuclei again. Studies of the intergalactic gas show that reionization was complete by $a \simeq 0.15$. The full cause of reionization is not well understood, but presumably it was in large part due to large amounts of ultraviolet radiation being produced from the just-forming galaxies. However, the free-electron densities in the intergalactic medium were low enough during reionization that even ionized gas was transparent.

Therefore, the upshot is that recombination first made the Universe transparent, and it has essentially remained so since that time. As a result, the CMB photons that we observe today have basically been travelling unhindered since recombination, so this event is also referred to as *last scattering.* We can observe the Universe back to this time, or equivalently back to this distance, which is also known as the *surface of last scattering.* Let us now calculate when that was.

A good estimate of when the epoch of recombination occured comes from considering the equilibrium of ionized and atomic hydrogen. In a fully atomic gas, the number density of atoms would equal that of baryons. But if a fraction χ of atoms is ionized, there would free electrons and protons in equal number density: $n_e = n_p = \chi n_{\mathrm{b}}$. The remaining atoms would be un-ionized atoms and have number density $n_{\mathrm{H}} = (1-\chi)n_{\mathrm{b}}$. Consider the dimensionless expression that combines all these terms:

$$\frac{n_{\mathrm{b}}\, n_{\mathrm{H}}}{n_e\, n_p} = \frac{1-\chi}{\chi^2}\,. \tag{9.10}$$

The number densities are all grouped together, and the right-hand side contains only the fraction χ. So, if we can compute the left-hand side, then we can solve for the ionization fraction. We now consider each of n_e, n_p, and n_{H} as classical ideal gases. Recalling the number density of such a gas from Eq. (4.31), we assume that the rest mass accounts for the chemical potential (that is, we put $\mu = 0$ in that formula) and drop the degeneracy for this approximate treatment. This leads to a number density

$$n_e \approx \left(\frac{m_e T}{2\pi}\right)^{3/2} e^{-m_e/T} \tag{9.11}$$

for electrons, and similarly fashioned ones for n_{H} and n_p. Substituting these densities into Eq. (9.10) produces

$$\frac{1-\chi}{\chi^2} \approx \eta\, n_\gamma \left(\frac{2\pi}{m_e T}\right)^{3/2} e^{(m_p+m_e-m_{\mathrm{H}})/T}, \tag{9.12}$$

where η is given by Eq. (9.4) and n_γ by Eq. (9.3). The mass difference $m_p + m_e - m_{\mathrm{H}}$ is not zero, and it represents the energy difference between ionized (free p^+ and e^-) and atomic (p^+ and e^- bound together) hydrogen; it is known as the ionization energy, and it is also given by $\frac{1}{2}m_e\alpha^2$.

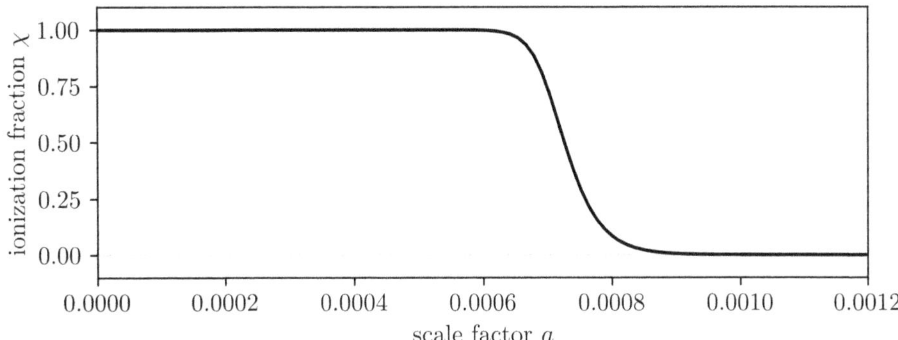

Fig. 9.1 The ionization fraction history of the early Universe. The time of recombination occured when the ionization fraction dropped and atomic hydrogen formed, turning the Universe transparent to radiation, which is observed today as the CMB. The time of large χ immediately preceding recombination is known as the last scattering.

Figure 9.1 plots the ionization fraction χ from Eq. (9.12) as a function of redshift. As we can see, $\chi \approx 1$ (complete ionization) at early epochs, but the fraction declines to $\chi < 0.01$ by $a = 0.0009$. Thus,

$$a_{\rm cmb}^{-1} \simeq 1100\,. \tag{9.13}$$

We see that recombination happened after the point of matter-radiation energy equality but before baryon–radiation equality (which had been calculated using the relation in Eq. (9.5)).

What happened between recombination at $a^{-1} = 1100$ and the earliest known galaxies at $a^{-1} \simeq 10$ is unexplored territory, sometimes called the dark ages. If you are thinking of doing research in astrophysics and looking for a seriously hard challenge, illuminating the dark ages is a good one. Meanwhile, look out for the new generation of telescopes, such as the Square Kilometer Array (SKA) radio telescope being built in South Africa and neighbouring countries.

Exercise 9.2 Compute the age of the Universe in years at recombination, in the concordance model with parameters in Eq. (8.18).

9.3 CMB Fluctuations

The conclusion from the previous section is that when we observe the microwave background sky, we are really seeing the recombination region that occured at $a_{\rm cmb}^{-1} \simeq 1100$. This is far, far 'behind' all the stars and galaxies (and hence the name, cosmic microwave *background*). This recombination region, the surface of last scattering, is remarkably uniform wherever we look across the sky. But it does have slight bumps and un-evenness at the level of 10^{-5}. These faint wrinkles, known as CMB fluctuations, are a snapshot of acoustic (i.e., pressure) waves in the photon–baryon plasma, seen at the epoch of recombination. These tiny fluctuations were discovered in 1992, by G. Smoot and colleagues with the DMR instrument on the COBE satellite, and have since been explored by several other microwave telescopes. The detailed theory for the plasma at recombination infers that the acoustic waves would have originated earlier, even closer in time to the Big Bang itself. The typical scale of the fluctuations is predicted to be the distance a pressure wave in the photon–matter plasma would have travelled since the Big Bang. This distance is called the sound horizon.

Now, one property of dark matter is that it does not carry the sound waves, since it interacts only gravitationally. So it is really a plasma of photons and baryonic matter involved, and at $a_{\rm cmb}$ photons are still somewhat dominant. The speed of sound in a photon-dominated plasma is $1/\sqrt{3}$ of the speed of light. That is, the sound horizon is $1/\sqrt{3}$ of the light horizon.

We cannot delve much more into the theory, but we can calculate the horizon size at recombination here. To do so we need to integrate the differential equation (8.20) relating r and a forwards from the Big Bang ($a = 0$). The result is

$$\Delta_{\rm hor} = a_{\rm cmb} \int_0^{a_{\rm cmb}} \frac{da}{a^2 H(a)} \,. \tag{9.14}$$

Here the integral gives the comoving distance that a light ray moves from the Big Bang to recombination, and multiplying by $a_{\rm cmb}$ leads to the physical size of the horizon. To work out how large this appears on the sky, we calculate the angular-diameter distance:

$$D_{\rm ang}(a_{\rm cmb}) = a_{\rm cmb} \int_{a_{\rm cmb}}^{1} \frac{da}{a^2 H(a)} \,. \tag{9.15}$$

The angular size of the horizon is

$$\theta_{\rm hor} = \frac{\Delta_{\rm hor}}{D_{\rm ang}} \,. \tag{9.16}$$

In general, the last two integrals must be computed numerically, using (8.17). For the Einstein-de Sitter case, however, the integrals simplify, and we have

$$\theta_{\rm hor} = \frac{\sqrt{a_{\rm cmb}}}{1 - \sqrt{a_{\rm cmb}}} \simeq \sqrt{a_{\rm cmb}} \,. \tag{9.17}$$

For the concordance values (8.18), $\theta_{\rm hor}$ is not very different. The sound horizon, then, is

$$\theta_{\rm hor}^{\rm sound} \simeq \sqrt{a_{\rm cmb}/3} \simeq 1^\circ \,. \tag{9.18}$$

This is indeed the characteristic size for the observed CMB fluctuations. You can have a look at them on APOD 050925 and APOD 130325. That random pattern goes back to the first moments of the Universe.

Appendix A
Rotations in Three Dimensions

Two Cartesian coordinate systems with a common origin may nonetheless differ in their orientation. They can be brought into alignment by a suitable rotation, which can be specified in several ways. One common convention in astronomy is to compose three rotations, as follows: first rotate about the z axis by ω, then about the current x axis by I, and finally about the current z axis (which has changed from the original z axis) by Ω. In terms of rotation matrices, we have

$$\begin{pmatrix} x & y & z \end{pmatrix}_{\rm rot} = \begin{pmatrix} x & y & z \end{pmatrix} R_z(\omega)\, R_x(I)\, R_z(\Omega)\,, \tag{A.1}$$

where

$$R_z(\omega) = \begin{pmatrix} \cos\omega & \sin\omega & 0 \\ -\sin\omega & \cos\omega & 0 \\ 0 & 0 & 1 \end{pmatrix} \tag{A.2}$$

and

$$R_x(I) = \begin{pmatrix} 1 & 0 & 0 \\ 0 & \cos I & \sin I \\ 0 & -\sin I & \cos I \end{pmatrix}\,, \tag{A.3}$$

while $R_z(\Omega)$ has the same form as $R_z(\omega)$.

We have used row vectors for the coordinates here. To change to column vectors, one can simply transpose the whole of Eq. (A.1). The rotation matrices get transposed and their order reverses, while the order of the rotations themselves remains the same:

$$\begin{pmatrix} x \\ y \\ z \end{pmatrix}_{\rm rot} = R_z^T(\Omega)\, R_x^T(I)\, R_z^T(\omega) \begin{pmatrix} x \\ y \\ z \end{pmatrix}. \tag{A.4}$$

Appendix B
Hamiltonians

Laws of physics can often be formulated as variational principles. The earliest example is Fermat's principle, which states that light takes paths for which the travel time is stationary compared to neighbouring paths. A light path between two points is typically a minimum of the light travel time, but it can also be a maximum, or even like a saddle point.

Hamilton's principle is a generalization of Fermat's principle. The version of it that we use in this book, for the chapters on celestial mechanics and Schwarzschild's spacetime, is covered in books on classical mechanics as Hamiltonian dynamics. It involves two sets of variables $\mathbf{p}$ and $\mathbf{q}$, known as the *canonical coordinates*, and time t. The $\mathbf{q}$ are coordinates and the $\mathbf{p}$ have the interpretation of momentum associated with each of those coordinates. The variables could be Cartesian

$$\mathbf{q} = (x, y, z), \quad \mathbf{p} = (p_x, p_y, p_z), \tag{B.1}$$

or polar

$$\mathbf{q} = (r, \theta, \phi), \quad \mathbf{p} = (p_r, p_\theta, p_\phi), \tag{B.2}$$

or more general things. The dynamics is given by a function $H(\mathbf{q}, \mathbf{p}, t)$ known as the Hamiltonian. Hamilton's principle is the variational principle

$$\delta \int \mathbf{p} \cdot d\mathbf{q} - H(\mathbf{q}, \mathbf{p}, t)\, dt = 0 \tag{B.3}$$

with $\mathbf{q}, t$ (but not $\mathbf{p}$) fixed at the boundaries of the integral. From the calculus of variations, Hamilton's principle implies that

$$\frac{d\mathbf{q}}{dt} = \frac{\partial H}{\partial \mathbf{p}}, \quad \frac{d\mathbf{p}}{dt} = -\frac{\partial H}{\partial \mathbf{q}}, \tag{B.4}$$

known as Hamilton's equations.

A simple but important Hamiltonian has Cartesian variables and Hamiltonian

$$H(\mathbf{r}, \mathbf{p}) = \tfrac{1}{2}\mathbf{p} \cdot \mathbf{p} + V(\mathbf{r}). \tag{B.5}$$

Hamilton's equations (B.4) are then

$$\dot{\mathbf{r}} = \mathbf{p}, \qquad \dot{\mathbf{p}} = -\nabla V(\mathbf{r}). \tag{B.6}$$

We recognize the equations for a particle of unit mass in a force field $-\nabla V(\mathbf{r})$. We can also see that H turns out to equal the total energy. (We have assumed unit mass.)

The usefulness of the Hamiltonian formulation is that it links global properties like symmetries to the differential equations for trajectories. For instance, if the function $H(\mathbf{p}, \mathbf{q})$ has no time dependence, it is constant along trajectories. This property is verifiable using Hamilton's equations (B.4). Other properties are difficult to derive from the equations, but almost obvious from the variational form. An important example is how the differential equations transform under coordinate transformations. Let us take this up in more detail.

Let $\mathbf{Q}, \mathbf{P}$ be a new set of variables, and let be $H(\mathbf{Q}, \mathbf{P}, t)$ be Hamiltonian such that dynamics is the same as under a previously-given $H(\mathbf{q}, \mathbf{p}, t)$. (We are using $H(\mathbf{q}, \mathbf{p}, t)$ and $H(\mathbf{Q}, \mathbf{P}, t)$ to denote two different functions, letting the arguments specify which function we mean.) The condition for the dynamics to be the same is complicated in terms of the differential equations, but in terms of the variational principle it is simple: the integrand in Hamilton's principle must not change, except possibly by a path-independent term. That is to say:

$$\mathbf{P} \cdot \dot{\mathbf{Q}} - H(\mathbf{Q}, \mathbf{P}, t) = \mathbf{p} \cdot \dot{\mathbf{q}} - H(\mathbf{q}, \mathbf{p}, t) + dS(\mathbf{Q}, \mathbf{p}, t)\,, \tag{B.7}$$

where $S(\mathbf{Q}, \mathbf{p}, t)$ can be any function of a mixture of new $\mathbf{Q}$ and the old $\mathbf{p}$. Its differential contributes a path-independent term. Further differentials can also be added. Adding $d(\mathbf{q} \cdot \mathbf{p})$ on the left, expanding the differentials, and comparing coefficients gives

$$\mathbf{q} = \frac{\partial S}{\partial \mathbf{p}}\,, \quad \mathbf{P} = \frac{\partial S}{\partial \mathbf{Q}}\,, \quad H(\mathbf{q}, \mathbf{p}, t) = H(\mathbf{Q}, \mathbf{P}, t) + \frac{\partial S}{\partial t}\,. \tag{B.8}$$

The transformation $(\mathbf{q}, \mathbf{p}) \to (\mathbf{Q}, \mathbf{P})$ is specified implicitly. It is called a canonical transformation and $S(\mathbf{Q}, \mathbf{p}, t)$ is a generating function for it. In this book, we use two simple but important examples of canonical transformations.

For orbits in Schwarzschild's spacetime we need to relate Cartesian and polar variables. Accordingly, let us identify $(\mathbf{q}, \mathbf{p})$ with (x, y, p_x, p_y), and take $(\mathbf{Q}, \mathbf{P})$ to be (r, ϕ, p_r, p_ϕ). We want

$$x = r\cos\phi\,, \quad y = r\sin\phi\,, \tag{B.9}$$

but how are p_r, p_ϕ related to the Cartesian variables? Let us set

$$S = (r\cos\phi)\, p_x + (r\sin\phi)\, p_y\,. \tag{B.10}$$

Substituting into the formula (B.8) for canonical transformations reproduces (B.9) as desired, and also supplies the accompanying relation for momentum variables

$$p_r = \frac{x p_x + y p_y}{r}\,, \quad p_\phi = x p_y - y p_x\,, \tag{B.11}$$

which we recognize as radial and angular momentum in two dimensions.

For the restricted three-body problem, we need to change to rotating Cartesian coordinates. That is, we want to relate $H(x, y, p_x, p_y)$ and (X, Y, P_X, P_Y) related by

$$\begin{pmatrix} x \\ y \end{pmatrix} \mathbf{M} \begin{pmatrix} X \\ Y \end{pmatrix}, \tag{B.12}$$

where

$$\mathbf{M} \equiv \begin{pmatrix} \cos t & -\sin t \\ \sin t & \cos t \end{pmatrix} \tag{B.13}$$

is the rotation matrix for unit angular velocity. Note that the inverse of $\mathbf{M}$ is just its transpose. The generating function

$$S = (p_x \; p_y)\, \mathbf{M} \begin{pmatrix} X \\ Y \end{pmatrix} \tag{B.14}$$

will do the job. Using the canonical transformation formula (B.8) again, we recover the rotating coordinates (B.12) again, and

$$\begin{pmatrix} p_x \\ p_y \end{pmatrix} = \mathbf{M}^{-1} \begin{pmatrix} P_x \\ P_y \end{pmatrix} \tag{B.15}$$

for the momentum components. The last equation in the canonical transformations formula (B.8) gives

$$\begin{aligned} H(x, y, p_x, p_y) &= H(X, Y, P_X, P_Y) \\ &\quad + (p_x \; p_y)\, \frac{\partial \mathbf{M}}{\partial t}\, \mathbf{M}^{-1} \begin{pmatrix} x \\ y \end{pmatrix}, \end{aligned} \tag{B.16}$$

relating the Hamiltonians. The last expression, despite its apparent complexity, simplifies to $xp_y - yp_x$. This we recognize as the angular momentum, which leaves us with a simple prescription: to change to a rotating frame (with unit spin) just add the angular momentum to the Hamiltonian.

Appendix C
Moving from Newtonian to Relativistic Frameworks

When moving from Newtonian (classical) to Einsteinian (relativistic) physics, several fundamental changes occur. For example, we move from a framework of three coordinate dimensions to a 4D one. Additionally, the role of 'time' changes in calculations, and we even have to introduce a new time variable. That being said, the Newtonian description is still a very useful limiting case of relativity, particularly on human scales and in daily life. Therefore, it is useful to understand both the differences and the connections between the systems. Here, we briefly discuss these changes in frameworks and their meaning. We start with a simple motivating example to build an intuitive sense of moving between coordinate systems, and then expand this to the relevant physical descriptions.

Consider the following example (from just a Newtonian point of view). A car moves along a line x with a constant speed $v = |dx/dt|$. This car is driven for some time interval T before running out of gas, which would correspond to a distance vT. Say that the car is also monitored with censors in the ground to check its progress over the x-axis, for upkeep purposes. When living in a 1D world on the x-axis, the car's driver agrees with the monitor's recorded distance.

Next, consider this car moving over a 2D plane while being constrained to the same constant speed v, which could now be split over two spatial coordinates, so that in general $v = \sqrt{(dx/dt)^2 + (dy/dt)^2}$. When driving along a trajectory *only* along the x-axis, both the driver and anyone reading the monitor output after a time interval T would agree that the car had travelled a distance vT. However, if the car's straight-line trajectory deviated from the x-axis, what would happen? The car's total speed is still v, but the speed component just along the x-axis, which is what the monitor measures, would be $v' < v$. The car would see its own distance travelled as vT, but the monitor would have recorded a shorter distance $\Delta x' = v'T$. Note that $\Delta x'$ would appear contracted compared to the car's $\Delta x = vT$ value because dx/dt and dy/dt additively combine (as the square root of squares) in the 2D velocity combination: having a component along the y-axis necessarily reduces that along the other coordinate axis here.

This distance mismatch occurs because the car is moving with a given velocity through more dimensions than the monitor is keeping track of: the car would describe its motion with a vector (x, y), and not just (x). According to the driver, the lower velocity in the x-direction is compensated by having a nonzero velocity in the y-direction. If the car's motion is mostly aligned with the x-axis, then this discrepancy would be very small and possibly hard to detect. But even that would only be a temporary relief—if we redrew the coordinate grid, then the dispute could be large, or at least different, and it seems odd that measures would depend so much on something as arbitrary as how coordinate axes were drawn. Note also that while time t appears as an interval and in the derivatives, it is not involved in velocity component trade-off—t is just a parameter, not a coordinate quantity like x and y.

The difference in the monitor and driver frameworks is similar to the basic difference in physics setups. The Newtonian framework considers motion through three dimensions (x, y, z), among which the 'trade-off' in having a constant velocity would occur, and 'time' t is a separate parameter used to schedule measurements ('position' is defined by the three coordinate values at a given time; derivatives of the coordinates are taken with respect to time; etc.). However, in the relativistic framework this is not the case, and what we commonly refer to as 'time' in Newtonian physics is actually another coordinate lumped in with the three spatial ones, so we must keep track of a 4D vector (t, x, y, z). When arguing about things moving at a constant speed, one would include 't' itself in with the trade-offs; thus, how one's velocity through x, y, and z changed would affect intervals measured in t. Since t is now one of the relativistic coordinate variables, we must use a new quantity to play the role of the independent parameter (i.e., what t was in the Newtonian case), called 'proper time', which is usually written as τ; in relativity, derivatives are made with respect to τ. This parameter τ describes the time measured in the car (so the driver in the vehicle considers it 'proper'), and an observer standing still in the frame coordinates would measure time with t (the definition of it being a coordinate quantity).

With these definitions and variables in place for the relativistic context, the driver and monitors can relate their measured quantities. Both the driver and the monitors will agree on the interval ds along with the driver moves. This quantity is known as the *Lorentz invariant*, and an observer in *any* inertial reference frame (i.e., moving at constant velocity) will measure the same interval ds. The driver will measure this 'proper distance' for the car in terms of the proper time τ and the constant speed of light c:

$$ds^2 = -c^2\, d\tau^2\,. \tag{C.1}$$

The stationary monitors in the road will measure this interval in terms of the coordinates to be:

$$ds^2 = -c^2\, dt^2 + dx^2 + dy^2 + dy^2\,. \tag{C.2}$$

Let us connect these quantities with the earlier car example for comparison and interpretation. If we try to construct a velocity $ds/d\tau$ in analogy to what we examined for the car example above, we see from Eq. (C.1) that this itself is an invariant quantity, so that:

$$\left(\frac{ds}{d\tau}\right)^2 = -c^2 = -c^2\left(\frac{dt}{d\tau}\right)^2 + \left(\frac{dx}{d\tau}\right)^2 + \left(\frac{dy}{d\tau}\right)^2 + \left(\frac{dz}{d\tau}\right)^2. \tag{C.3}$$

First, looking at the right-hand side, we note that a term with t is, indeed, included in parallel with the spatial coordinate quantities. However, t itself still retains some special features and differentiability from the spatial coordinates: the time term bears a coefficient of c^2, and it also has a sign difference.[1] The units of the coefficient c permit each term to be additively combined, but more meaningfully, its magnitude $\approx 3 \times 10^8$ m s^{-1} means that it takes *very* large, quick changes of spatial coordinates (on human scales) for offsetting changes in t to become apparent. The fact that the weighting is so large relative to the spatial velocities we normally experience on Earth is part of the reason why Newtonian approximations work so well for most everyday activities—typically, they are analogous to the case in which the car moves almost exactly along the x-axis and the other effects are tiny. The sign difference in the t-term affects how the trade-offs among the coordinates occur:[2] e.g., as spatial velocity component grows, time would appear to dilate. We note that since the middle quantity of $-c^2$ is a constant, any changes in spatial coordinates *must* result in a change in the time coordinate, and vice versa.

In the 2D car example we saw that leaving out coordinates could result in disputes about velocity and distance intervals. Similarly (and symmetrically), there would be disputes about *time* intervals if a full set of coordinates is not in place. By equating the Lorentz invariant definitions in Eqs. (C.1) and (C.2), we can relate the intervals of proper time τ and coordinate time t (by which the driver and observer, respectively, record measurements) by:

$$d\tau^2 = dt^2 - \frac{1}{c^2}\left(dx^2 + dy^2 + dz^2\right) . \tag{C.4}$$

Here again, the time t is locked together with the spatial coordinates (with an appropriate weighting factor, c^{-2}). Thus, a changes in any of the spatial coordinates results in a relative *increase* of the coordinate time relative to proper time (which was also noted above, as a property of its sign difference).

In summary this example and discussion has been aimed at describing a fundamental conceptual feature of relativity, and one that is a principle difference from the Newtonian model. The Newtonian framework has three coordinates, and time as a separate parameter; the relativistic framework is inherently 4D, incorporating time as a coordinate. As a consequence, we cannot separate moving through space and making measurements of time– they are intimately connected as being part of the same 4D-quantity. Instead of talking about position (x, y, z) at a given time t, in relativity we have the *event* (t, x, y, z) and reference it with proper time parameter τ. The importance of including time as a coordinate becomes vital when dealing with fast, high energy motions.[3]

[1] Additionally, besides the weighting and sign difference, on an experiential level we note that the time coordinate behaves differently than the spatial ones: while we can move back and forth freely in space, revisiting places, we seem to progress along the time axis in just one, unchanging direction. The *reason* for that feature is unknown, and it is an aspect of the coordinates that is outside of the principles of relativity. It is 'merely' a fact of the way things appear to be, and perhaps something for philosophical debate. It is likely that the several distinctions that t has from the spatial coordinates led to it being considered an entirely separate entity for so long before the development of relativity.

[2] We also note that in the initial car example above, we *chose* to make the velocity constant—in the relativistic case, the proper time derivative of the Lorentz invariant $ds/d\tau$ is constant *by definition.*

[3] The relations in Eqs. (C.1) and (C.4) were described for the case of a moving, massive particle—namely, the car and driver. The equations likewise apply to photons, which are massless, but the Lorentz invariant is then $ds = 0$.

More broadly, by incorporating this 4D viewpoint, disputes about measured quantities of time, distance, and velocity can be avoided. Einstein, Poincaré, and others first used this new coordinate basis to generate a framework for physical laws based on the *special* principle of relativity, that all laws appear the same in any *inertial* frame (i.e., one moving at constant velocity with respect to another, as in the example of the car driving on the grid above). This framework was soon extended to establish one following the *general* principle of relativity, which applied this equivalence to observers in *any* frame: both being stationary, one moving steadily, one in an accelerating frame, one navigating near a black hole, etc.

Appendix D
Working with Planckian Units

Planckian units are wonderful for simplifying formulas and revealing their physical meaning. But for Planckian units to be useful, one needs to be able to convert to and from SI units whenever needed. Chapter 4 explains how to do such conversions. This appendix just tabulates the relevant definitions and values used through the book.

The essential quantities are the Planckian units themselves, as given in Eq. (4.3):

$$\begin{aligned} \text{marb} &\equiv \sqrt{\hbar c/G} &&= 2.177 \times 10^{-8}\,\text{kg}\,, \\ \text{lap} &\equiv \sqrt{\hbar G/c^3} &&= 1.615 \times 10^{-35}\,\text{m}\,, \\ \text{tick} &\equiv \sqrt{\hbar G/c^5} &&= 5.383 \times 10^{-44}\,\text{s}\,, \\ \text{therm} &\equiv \sqrt{\hbar c^5/G}\,k_{\text{B}}^{-1} &&= 1.419 \times 10^{32}\,\text{K}\,. \end{aligned} \tag{D.1}$$

The definitions are standard, but the units have no conventional names, so we have made up some names for this book.

The fundamental physical constants in this book, given in Planckian units in Eq. (4.2) and whose values compose the conversion factors in Eqs. (4.3) and (D.1), are

$$\begin{array}{llll} c & = 2.99792458 & \times 10^{8} & \text{m s}^{-1}\,, \\ G & = 6.674 & \times 10^{-11} & \text{m}^3\ \text{kg}^{-1}\ \text{s}^{-2}\,, \\ 2\pi\hbar & = 6.62607004 & \times 10^{-34} & \text{J s}\,, \\ k_{\text{B}} & = 1.38064852 & \times 10^{-23} & \text{J K}^{-1}\,. \end{array} \tag{D.2}$$

Only three microphysical parameters are used in this book: the electron mass, baryon mass (taken as the mean of proton and neutron masses), and the electromagnetic coupling (fine structure) constant. The values of these in terms of Planckian units are:

$$\begin{array}{lll} m_e & = 4.184 \times 10^{-23} & [\times\,\text{marb}]\,, \\ m_{\text{b}} & = 7.688 \times 10^{-20} & [\times\,\text{marb}]\,, \\ \alpha & = 1/137.036\,. & \end{array} \tag{D.3}$$

Several characteristic scales are defined in terms of the above parameters, of which the most important are the Bohr radius, the Thomson cross-section, and the Landau mass:

$$\begin{array}{lll} r_{\text{B}} & = (\alpha m_e)^{-1} & [\times\,\text{marb}\,\text{lap}]\,, \\ \sigma_{\text{T}} & = (8\pi/3)\,(\alpha/m_e)^2 & [\times\,\text{marb}^2\,\text{lap}^2]\,, \\ M_{\text{L}} & = m_{\text{b}}^{-2} & [\times\,\text{marb}^3]\,. \end{array} \tag{D.4}$$

When reconstructing the SI version of a formula, it is necessary to include the dimensions ([× marb] and so on). But to calculate values in SI units—as explained in

Chapter 4—only the numerical values are needed, and the dimensions can be restored at the very end of a computation.

Some astrophysical quantities, such as the solar and Earth masses, can be respectively expressed in microphysical terms as:

$$\begin{aligned} M_\odot &= 0.54\, M_{\mathrm{L}}\,, \\ M_\oplus &= 1.5 \times 10^{-6}\, M_{\mathrm{L}}\,. \end{aligned} \tag{D.5}$$

Others quantities are consequences of natural history and have no microphysical interpretation:

$$\begin{aligned} \mathrm{au} &= 499.00\,\mathrm{s}\,, \\ \mathrm{parsec} &= 1.029 \times 10^{8}\,\mathrm{s}\,, \\ {H_0}^{-1} &= 14 \pm 0.5\,\mathrm{Gyr}\,, \\ T_{\mathrm{cmb}} &= 2.725\,\mathrm{K}\,. \end{aligned} \tag{D.6}$$

Where these are used in Planckian formulas, it is assumed that they have been converted explicitly into Planckian units.

References

Binney, J. (2016). *Astrophysics: A Very Short Introduction.* Oxford University Press.

Binney, J., and Tremaine, S. (2008). *Galactic Dynamics,* 2nd edn. Princeton University Press.

Byers, N., and Williams, G. A. (2006). *Out of the Shadows: Contributions of Twentieth-Century Women to Physics.* Cambridge University Press.

Carroll, B. W., and Ostlie, D. A. (2006). *An Introduction to Modern Astrophysics and Cosmology.* Pearson, Addison-Wesley.

Carter, B. (2007). The significance of numerical coincidences in nature. *ArXiv 0710.3543.*

Chandrasekhar, S. (1939). *An Introduction to the Study of Stellar Structure.* University of Chicago Press.

Chandrasekhar, S. (1995). *Newton's Principia for the Common Reader.* Oxford University Press.

Choudhuri, A. R. (2010). *Astrophysics for Physicists.* Cambridge University Press.

Christensen, N., and Moore, T. (2012). Teaching general relativity to undergraduates. *Physics Today*, **65**(6), 41.

Clayton, D. D. (1984). *Principles of Stellar Evolution and Nucleosynthesis.* University of Chicago Press.

Clayton, D. D. (1986). Solar structure without computers. *American Journal of Physics*, **54**, 354–62.

Diacu, F., and Holmes, P. (1996). *Celestial Encounters: The Origins of Chaos and Stability.* Princeton University Press.

Kennefick, D. (2012). Not only because of theory: Dyson, Eddington, and the competing myths of the 1919 eclipse expedition. In *Einstein and the Changing Worldviews of Physics* (ed. C. Lehner, J. Renn, and M. Schemmel), pp. 201–32. Birkhäuser Boston.

Linder, E. V. (1997). *First Principles of Cosmology.* Prentice Hall.

Maoz, D. (2016). *Astrophysics in a Nutshell,* 2nd edn. Princeton University Press.

Morais, M. H. M., and Namouni, F. (2016). A numerical investigation of coorbital stability and libration in three dimensions. *Celestial Mechanics and Dynamical Astronomy*, **125**, 91–106.

Padmanabhan, T. (1996). *Cosmology and Astrophysics through Problems.* Cambridge University Press.

Padmanabhan, T. (2006). *An Invitation to Astrophysics.* World Scientific.

Ryden, B. (2016). *Introduction to Cosmology,* 2nd edn. Cambridge University Press.

Shapiro, S. L., and Teukolsky, S. A. (1986). *Black Holes, White Dwarfs and Neutron Stars: The Physics of Compact Objects.* Wiley.

Taylor, D. B. (1981). Horseshoe periodic orbits in the restricted problem of three bodies for a sun–Jupiter mass ratio. *Astronomy and Astrophysics*, **103**, 288–94.

Wheeler, J. A. (1998). *Geons, Black Holes and Quantum Foam: A Life in Physics.* W. W. Norton.

Index

www.ingramcontent.com/pod-product-compliance
Ingram Content Group UK Ltd.
Pitfield, Milton Keynes, MK11 3LW, UK
UKHW051126260726
13967UKWH00010B/2894

9 780198 816478